SAGE was founded in 1965 by Sara Miller McCune to support the dissemination of usable knowledge by publishing innovative and high-quality research and teaching content. Today, we publish more than 750 journals, including those of more than 300 learned societies, more than 800 new books per year, and a growing range of library products including archives, data, case studies, reports, conference highlights, and video. SAGE remains majority-owned by our founder, and after Sara's lifetime will become owned by a charitable trust that secures our continued independence.

Los Angeles | London | Washington DC | New Delhi | Singapore | Boston

Perspectives on India's Defence Offset Policy

Thank you for choosing a SAGE product! If you have any comment, observation or feedback, I would like to personally hear from you. Please write to me at contactceo@sagepub.in

—Vivek Mehra, Managing Director and CEO,
SAGE Publications India Pvt Ltd, New Delhi

Bulk Sales

SAGE India offers special discounts for purchase of books in bulk. We also make available special imprints and excerpts from our books on demand.

For orders and enquiries, write to us at

Marketing Department
SAGE Publications India Pvt Ltd
B1/I-1, Mohan Cooperative Industrial Area
Mathura Road, Post Bag 7
New Delhi 110044, India
E-mail us at marketing@sagepub.in

Get to know more about SAGE, be invited to SAGE events, get on our mailing list. Write today to marketing@sagepub.in

This book is also available as an e-book.

Perspectives on India's Defence Offset Policy

Edited by
ManMohan S. Sodhi
Rajiv Bhargava

www.sagepublications.com
Los Angeles • London • New Delhi • Singapore • Washington DC • Boston

First published in 2015 by

SAGE Publications India Pvt Ltd
B1/I-1 Mohan Cooperative Industrial Area
Mathura Road, New Delhi 110 044, India
www.sagepub.in

SAGE Publications Inc
2455 Teller Road
Thousand Oaks, California 91320, USA

SAGE Publications Ltd
1 Oliver's Yard, 55 City Road
London EC1Y 1SP, United Kingdom

SAGE Publications Asia-Pacific Pte Ltd
3 Church Street
#10-04 Samsung Hub
Singapore 049483

Published by Vivek Mehra for SAGE Publications India Pvt Ltd, typeset in 11/13 Minion Pro by RECTO Graphics, Delhi and printed at Sai Print-o-Pack New Delhi.

Library of Congress Cataloging-in-Publication Data Available

ISBN: 978-93-515-0139-8 (HB)

The SAGE Team: Sachin Sharma, Alekha Chandra Jena and Vaibhav Bansal

Contents

List of Tables

Foreword

India has finally managed to frame an offset policy after much iteration, debate and delay. The question now being asked is: how do you actually make it work effectively on ground? Clearly, what has happened so far does not quite reflect the potential and possibility.

Given our demographics, our population and our state of development, we have no choice but to enhance manufacturing and increase the number of jobs on offer. Offsets provide such an opportunity.

India needs to take a leaf out of the Japanese experience. While creating their manufacturing ecosystem, the Japanese took a good hard look at what was happening in different parts of the world; where the best engineering was happening, where the best technology was being developed and where the best products were available. Japanese manufacturing firms and research institutes then tried to make these products quicker, better and cheaper.

Offset deals with Western nations provide a unique opportunity for the Indian manufacturing sector to grow and acquire competencies. Either we could look at getting all the technology from outside—where foreign companies come and manufacture here— or we could pick specific areas for developing Indian competence. Or perhaps we could opt for a combination of approaches, and focus on verticals and segments where Indian firms will be able to innovate and adapt the fastest.

A successful offset policy will work better in India if we strengthen the research and development (R&D) ecosystem in India. For many decades, innovation and research were not rewarded enough in India; nor were institutional–corporate linkages encouraged. Tax rates, right up to the 1990s, were so high that there was very little surplus left for research. Today, policies are more favourable with weighted deductions provided on R&D investments, yet the mindset is still steeped in the past. Even though there is gradual change taking place now, it needs to be accelerated and become pervasive in India as the quality movement did two decades ago.

Strong encouragement and active execution of the offset policy are needed by the government to galvanize India's corporate sector. Perhaps there is even a case for the government to handpick organizations and sectors in which we could deliver and build leadership positions. We only need to study the experiences of nations such as Japan, Korea and China in this area to chalk out a time-bound road map and conducive incentive structures for ourselves, to build a number of world-class companies with leadership positions in their chosen fields.

To make a successful offset policy also means the different perspectives of the government, the foreign original equipment manufacturers (OEMs) and Indian industry are all taken into account to create win–win possibilities. I congratulate the editors, Prof. ManMohan S. Sodhi and Col. Rajiv Bhargava, for compiling these perspectives in a single volume, something that has never been done before and something that will serve policy makers and industry for years to come.

S.K. Munjal
Co-Chairman, Hero MotoCorp

Preface

The Defence Procurement Procedure (DPP) and defence offsets are evolving continually based on respective experiences of stakeholders in how they were implemented at a particular time. The DPP has come a long way since it was first introduced in 2001 and credit for that goes to the feedback from the users and the flexibility of the policy makers. However, there remain complexities in the policy and a perception of red tape against the background of unmet and unclear objectives.

To understand this landscape from the perception of different stakeholders, we set out to compile the widely divergent perspectives of the Indian government and analysts, of the foreign original equipment manufacturers (OEMs) and of Indian industrialists. The evolution in policy can be understood as a result of the three-way tussle of these stakeholders. Our reasoning was that putting all their viewpoints together would help each side at least understand the others' views, and possibly lead to a position that is better for all. Having all viewpoints in one place may also shed light on why policies have evolved thus far in the way they have and on what possible trajectories defence procurement may take in the future.

This book provides the views of various stakeholders in defence production and procurement on their understanding of the defence offset policy. For the first time ever, the government, analysts, industrialists and OEMs present their views on a common platform. The authors of the contributed chapters include top tiers of the Indian government and think tanks, senior managers of global OEMs and the leadership of Indian companies. Together, the chapters cover a wide range of issues: requirements of investment in R&D, requirements of supply chain management and financial issues that would make Indian industry competitive in this domain.

The reader will find how different stakeholders are thinking about such fundamental questions as:

- What are the objectives of the defence offset policy and what could be the objectives in the future if different from those today?
- What should or could be the defence export policy given the historic policy of not exporting weapons?
- What are the roadblocks to India having a manufacturing base catering not only to defence but to the economic improvement of the country?
- What would make India attractive as a destination for investment to foreign OEMs who would also bring in technology?

The purpose of this compilation is not to provide answers to these questions though—rather, our belief is that to find answers workable to all, all stakeholders must first listen to each other and then discuss. As such, if this book generates a constructive debate on Indian defence procurement and the defence offset policy among these stakeholders, the purpose of the book would be achieved.

ManMohan S. Sodhi
Rajiv Bhargava

PART I
AN OVERVIEW OF DIFFERENT
PERSPECTIVES

1

Divergent Views on Defence Offsets

ManMohan S. Sodhi and Rajiv Bhargava

There are divergent views on what offsets are, whether they can achieve the objectives they are designed to achieve, and on the best design assuming offsets work. Broadly we can consider the government, the foreign original equipment manufacturers (OEMs) and the domestic industry as stakeholders holding divergent views. But even within each category there is divergence, so defence offsets are controversial.

First let us consider the definitions. Offsets are formal trade arrangements when a foreign supplier undertakes specific programmes to fulfil an urgent economic need or to boost indigenous production through technology infusion.[1] Another definition is that offsets are in-country compensations required by the government as a condition for purchasing major goods and services.[2] Yet another definition is that an offset is 'a deduction; a counterclaim; a contrary claim or demand by which a given claim may be lessened or cancelled; a claim that serves to counterbalance or to compensate for another claim'.[3]

These definitions include objectives, but can the practice of offsets actually meet these objectives? Some economists believe that the practice is economically inefficient while those in government

---

[1] Mrinal Suman, Chapter 4.
[2] Yannis Ailianos, Chapter 13.
[3] Black's Law Dictionary 1085 (6th ed. 1990).

feel offsets are invaluable given the imperfections and asymmetries of the defence procurement market.

Transparency International dismisses offsets: 'Offsets are claimed to offer economic benefits to countries. However, the current economics literature reveals the weakness and uncertainty of this claim. Arranging offsets adds costs to the value of the defence purchasing contract, and these costs are borne by the importing country and not by the supplier'.[4]

Countering that is the view that countries need offsets given that the markets are imperfect and we need to deal with the OEM oligopoly. As one government official puts it, 'Defence markets are not free markets, they are narrow, with oligopolistic and market imperfections. Offsets are required because they seek to counter these distortions'.[5] Another official notes, 'In competitive markets, economic rents are supposed to be just competed away by new entry of firms. But in an imperfect market such as an oligopoly, factors such as a steep learning curve or high levels of investment, etc. curb the entry of new firms and cause economic rents to persist. Offsets strategy may be used to share in some of this economic rent'.[6]

DESIGNING AN OFFSET POLICY

Assuming offsets work, there are questions on how to best design policies on their use, given the objectives. Governments seek to design offset policies that suit their aspirations. Typically, an offset policy needs to be clear and realistic to account for domestic capabilities while addressing global perspectives, and needs to be embedded in a wider national policy. The policy objectives of a government are fivefold: (1) national priorities (e.g. self-reliance); (2) specific sector (e.g. the defence sector); (3) specific technologies

---

[4] Transparency International, 'Defence Offsets: Addressing the risks of corruption and raising transparency', April 2010.

[5] Vivek Rae, Speech at Swavalamban II Conference at ISB Hyderabad, 2001.

[6] Vijayalakshmy K. Gupta, Chapter 3.

(e.g. critical knowhow in sensors); (4) specific means to achieve national or other objectives (e.g. job creation) and (5) specific relevant activities (e.g. foreign investment).

If we think of any government policy as being designed around a set of parameters, the offset parameters chosen and set for meeting these objectives typically pertain to (a) regulations, (b) incentives and (c) monitoring mechanisms.

Taking this 'parameters' view, one OEM executive summarizes how offset policy is made from their perspective: 'First the government defines the policy objectives, strategic fit on why offsets, the aspirations, the policy goal, industrial goals, national priorities, etc. It defines long-term objectives of the policy. Once this is done, typically the authorities then determine the offset parameters: the areas eligible for offset, the volume, timeframe, etc. Obviously these parameters need to come in support of the long-term policy objectives. Now once these parameters [have been] actually defined the governments can issue tenders including offset requirements that the foreign contractors can respond to. But this is not all. Once the tenders have come into a contract, there is another important element that needs to be emphasized. It is for the government to establish a monitoring mechanism to make sure that the requirements have been achieved, not only to protect the national interest but also to support dual implementation of the offset schemes to enable the long-term policy objectives'.[7] In India, the defence offset policy can be broadly described using a number of parameters as shown in Table 1.1. Note the difference between values in India in comparison to typical values for other countries although each country is different.

The direction that the policy takes is determined by the way each of these parameters is tuned. Even minor incremental changes can impact policy disproportionately. Table 1.2 defines how changes (positive or negative) in the drivers impact economic considerations.

---

[7] Yannis Ailianos, Chapter 13.

Table 1.1   Comparison of parameters for offsets: India and others

Parameter	Current status in India	Global values
Threshold for offset initiation	US$65 million	US$1–100 million
Offset liability	30% (flexible)	110%
FDI cap for offsets	26% but moving to 49%	100%
Multipliers	1–3	1–15
Counter trade sector	Defence	Defence, Civil
Offset sector	Defence, Aerospace, Inland security	Defence, Civil, mixed
Banking provision	7 years	Varying
Time validity of offsets	2 years	Varying
Penalties	5% of unfulfilled contract, with ceiling at 20% of offset contract value	NA
Exports	Nil	100%
Offset partners	Indian industry, MSMEs	NA
Currency of transactions	Indian rupees	US$

Table 1.2   Impact of drivers on economic considerations

Modifying parameter	Pros	Cons
Enhancing offset liability	More flexibility in allocating multipliers	Costlier imports
Higher FDI in offset joint venture	Incentives to OEMs for transfer of technology (ToT), more commitment from OEM, higher reliability	Reduction in government control, higher say of OEM
Increasing multipliers	Tool to channelize ToT to priority sectors and priority partners	Methodology of fulfilment is key issue; valuation of technology is a challenge

(Table 1.2 Contd.)

(Table 1.2 Contd.)

Modifying parameter	Pros	Cons
Enhancing banking provision	Fosters long-term investments by OEMs; affords flexibility	Opportunity for misuse in terms of exchange of credit
Increasing time validity of offset	More flexibility to OEMs	Scope for delay in fulfilment of obligation
Enhancing indirect offsets through a National Offset Policy	All sectors benefitted; rising tide to raise all boats; market distortion reduced	Benefits to defence production diluted over short term; turf issues may gain prominence
Positive outlook on exports	More markets, economy of scale	No policy exists on exports, bureaucratic hurdles, ideological issues
Higher policy clarity	More transparency of process, focused investments, more value for money, reduced kickbacks	No evident cons
Stringent implementation of penalty clause	Timely fulfilment of offsets	Potential for misuse
More teeth to monitoring agency	Effective monitoring of commitments	Potential for misuse

THE STAKEHOLDERS

In shaping this book, and this comes out from the diverse perspectives in this book, there are three major stakeholders in the practice of defence offsets:

- Government (including the public sector undertakings)
- Foreign OEMs as offset donors
- The domestic private sector as offset recipients

Each set of stakeholders has its own objectives and approach towards defence offsets. The fourth category is that of the analysts who offer a nonpartisan opinion on issues in this complex practice; in this book, we have bundled their views with that of the government because the analysts have broadly presented the Indian viewpoint. For offsets to become win–win solutions, willing participation of all three as partners is a must.[8]

As offset parameters are used to determine the direction of the offset policy, each stakeholder seeks to tweak the parameters to further its own objectives. As a starting point, it would be useful to know how these stakeholders interpret policy and its preferred future direction from their viewpoint.

PERSPECTIVES OF THE GOVERNMENT AND ANALYSTS

It is clear to the Indian government that defence offsets are only a means to an end: developing the defence industrial base. 'Now there is increased realization that the goal of achieving self-reliance in indigenous development of defence systems cannot be achieved without enhancing the defence industrial base in the country and that would increase the production of indigenously developed systems. The only way to achieve this is by creating additional capacities in production, in private industries also, both large and small as system integrators for SMEs for niche systems'.[9]

Moreover, 'the role of the government is to be a facilitator, to create an environment, which sends out signals and creates incentives for enterprises to respond in ways that the government would like them to respond'.[10] The government is clear about the role of the private sector as a partner in expansion of the defence industrial base of the country. 'The government and the Ministry is deeply committed towards strengthening the Offset program,

---

[8] M.K. Badhwar, Chapter 17.

[9] E.S.L. Narasimhan, Chapter 2.

[10] M.M. Pallam Raju, Speech at Swavalamban II Conference at ISB Hyderabad, 2001.

we are deeply committed towards encouraging the participation of the private sector'.[11]

In wishing to work with the private sector, the Indian government has a healthy respect for the domestic industry. 'Indian Industry has developed a strong Industrial base with a successful track record of implementing technology-intensive projects including bulk production within stipulated time frames at reasonable cost and world class quality'.[12] 'Many of the Indian private sector industries have today moved up the value chain from build to print, build to specification, build to design and build to requirements. This has catapulted many industries from being supplier of products to becoming Defence Research & Development Organisations (DRDO) to eventually becoming supplier of products to multinationals'.[13]

Indian analysts are generally favourably disposed towards defence offsets. 'Being one of the biggest [importers] of defence equipment, India can draw immense benefits with a well-thought-out offset policy'.[14] Offsets will make a difference to the industry, if the industry makes business sense of the transaction involved.[15]

These perspectives can be under the following issues: (1) technology transfer, (2) direct and indirect offsets, (3) enhancement of foreign direct investment (FDI), (4) capability building, (5) the role of Defence Public Sector Undertakings (DPSUs) and (6) accountability of offsets.

Transfer of Technology

In the global arms business, availability and pricing of technology is not just a function of economic values. Political and strategic

---

[11] M.M. Pallam Raju, 2011. Speech at Swavalamban II Conference at ISB Hyderabad.

[12] Vijayalakshmy K. Gupta, Chapter 3.

[13] E.S.L. Narasimhan, Chapter 2.

[14] Mrinal Suman, Chapter 4.

[15] K.V. Kuber, Chapter 8.

considerations are often overriding factors, and certain critical technologies may not be even available for any price. As one analyst puts it, 'World trade for defence equipment is not run on economic considerations alone. Technology denials have been and will always be used as an instrument of foreign policy'.[16] When India sought critical space technology from the US in the 1990s, the MTCR and other strategic issues were of paramount importance and no amount of money could have enabled that purchase.

The general opinion amongst analysts is that India's buying capacity in an otherwise stagnating global arms market can be leveraged to obtain technology that is not available otherwise. A typical view is that 'with the kind of expenditure that we have globally more or less unmatched over the next two decades, India has a huge opportunity to translate its power as a large [buyer] into forcing through offsets the kind of indigenous capability that we desire'.[17]

The eventual goal is also seen as leading towards self-reliance while at the same time becoming part of the global value chain. India is one of the largest importers of defence equipment and has a policy of not exporting. So it is unclear how the country could be part of a global value chain. Still, both are seen as prioritised goals: 'the primary consideration for introduction of offsets in the Indian acquisition system was to leverage defence manufacturing base in the country and address the needs of self-reliance. The secondary consideration was to become part of the global supply chain thereby giving impetus to defence exports'.[18]

Technology has a critical role to play in achieving self-sufficiency in defence production. Availability of technology to Indian industry would enable it to transform companies from being, say, Tier-3 suppliers to becoming key Tier-1 suppliers to global OEMs and even global integrators in the long term. As an analyst puts it, 'the Indian Industry desperately needs technology to be able to come up to global manufacturing standards'.[19]

---

[16] R.S. Bhatia, Chapter 14.
[17] V. Sumantran, Chapter 15.
[18] R.S. Bhatia, Chapter 14.
[19] K.V. Kuber, Chapter 8.

There are four policy incentives by which global OEMs can be motivated to transfer technology to the offset receiver: (1) inclusion of technology transfer in offset offerings, (2) offering multipliers on buyback by OEMs, (3) relaxing the FDI limit and (4) increasing the share of indirect offsets in the entire pie. The Ministry of Defence (MoD) in its recent Defence Offset Guidelines has accepted a model worked out with assistance of DRDO, which addresses the issue of inclusion of transfer of technology (ToT) in offsets and the issue of multipliers to a fair extent. However, the issues of enhancing FDI and stepping up the share of indirect offsets have been partially addressed. After all, as one analyst acknowledges: the limitation of FDI in case of Indian offset partners to 26 per cent has been a sore point and needs due diligence to provide confidence to OEMs in transferring technologies.

ToT is one of the most common types of offset in the defence transactions; however, it is not among the simplest. One analyst explains, 'A successful ToT depends on four essentials—careful identification of technology for import, selection of [a] reliable OEM, objective nomination of recipient and close oversight of technology absorption. A truly successful ToT would enable the recipient to produce equipment from component and raw-material level'.[20]

Serious concerns regarding ToT include the degree of obsolescence, the evaluation of technology, documentation and configuration control and protection of intellectual property rights (IPR): 'When we talk about technology transfer, the experience over the globe has been that mostly obsolete technology has been transferred; only the degree of obsolescence varies'.[21] The evaluation and pricing of technology is another issue: 'There is no standard price of technology in the world market. It is purely need based and determined by the degree of desperation of the technology seeker'.[22] The design documents are generally not provided in entirety as the OEM is not willing to part with them. Hence, 'complex interface

---

[20] Mrinal Suman, Chapter 4.
[21] S.N. Misra, Chapter 5.
[22] Mrinal Suman, Chapter 4.

issues arise during the integration process for which solutions are not clearly defined in the ToT document'.[23] IPR and technology threats have resulted in losses of billions of dollars to IPR creators, in this case mostly OEMs. The view of OEMs that need to transfer technologies to Indian technology companies, IOPs and Indian Strategic Electronics (ISE) companies is coloured by their experiences on ToT with Japan, China, Singapore, South Africa, South Korea, Brazil and Israel. The recent issues on violation of IPRs and Alleged technology thefts in China have resulted in OEMs and governments viewing India through the same lens.

Since 2010, the government has put in effort to streamline the defence acquisition process. However, the results have been slow in terms of their outcomes. As a senior official notes, 'even today, all the major Defence deals are signed with foreign OEMs. The public sector continues to get bulk orders under transfer of imported technology. The private sector continues to be a peripheral participant with the production of some low-tech items'.[24]

There is overall consensus among government officials and analysts regarding technology transfer. 'The experience thus far has proved that technology cannot be bought off the shelf and such a buy will invariably be riddled with restrictions in terms of licensed manufacture with little scope for modifications or upgrades to the technology offered', according to a government official.[25] Another notes that 'technology transfer through offsets is an efficient method of technology acquisition because in the direct purchase of technology the entire risk is on the buyer but when technology is part of offsets then the risk also shifts to the seller because he has greater incentive to transfer the technology as it is linked to the acquisition of the entire system'.[26] Moreover, 'these are a way of infusing new technology into the economy, thereby ultimately stimulating economic growth. The argument is that the social benefits of technology induction exceed the individual benefit to

---

[23] S.N. Misra, Chapter 5.
[24] E.S.L. Narasimhan, Chapter 2.
[25] Ibid.
[26] Shobhana Joshi, Chapter 7.

the firm or industrial enterprise. Thus, the firm will purchase new technology only up to the point where its marginal benefit equals its marginal cost. Hence, the government may use industrial policy (offsets) as a way to induce higher levels of investment in technology which has the spread effect for the entire economy'.[27]

Another issue around ToT through offsets is to carry out the valuation of technology and the costing of offsets. As noted by one official, 'various methodologies are used to bridge the gap between the perspective of the buyer and seller. The buyer focuses on the value because offsets provide an opportunity to get the desired defence technology. For the seller, on the other hand, the cost incurred to implement the offsets assumes significance. For the buyer or the recipient, the bottom line is that the price of the equipment should be reasonable after the cost of the technology is amortized on the minimum order quantities'.[28]

Overall there is a sense that technology transfer needs to be revamped thoroughly: 'While looking at offsets, Government cannot stop at merely making Indian industry a part of global supply chain of defence majors but also to seek system integration in India.... The myth regarding capability of the Indian industry to absorb the huge quantum of defence offsets needs to be looked at from the consideration of capability and capacity and track record'.[29]

Direct and Indirect Offsets

There is much debate on the issue of direct versus indirect offsets. Offsets in India are mostly direct, with the indirect component (aerospace) having being added only recently. On the issue of enhancing the share of indirect offsets in the offsets pie, the government (including PSUs) generally opposes the idea while analysts and Indian industry are firmly in favour.

---

[27] Vijayalakshmy K. Gupta, Chapter 3.
[28] Shobhana Joshi, Chapter 7.
[29] Vijayalakshmy K. Gupta, Chapter 3.

Analysts generally feel that not only should indirect offsets have a larger share, they should form part of the National Offset Policy as that would benefit manufacturing in the entire country rather than just defence production. Some analysts also feel that offset obligations should be raised by 100 per cent, with 30 per cent going to defence production and the remaining 70 per cent going to other high tech sectors. '100 per cent offset obligations under defence offsets need to be set. While 30 per cent can continue to be allotted to the defence production sector, 70 per cent can be used in high-tech sectors such as aerospace, telecommunications, railways, composites, machine tools, electronics hardware, and other requirements of India's high-tech industry'.[30]

Likewise, in the view of another analyst, 'An important aspect in offset policy is the need of an effective mix of indirect and direct offsets. Globally, 60 per cent of the countries opt for both direct and indirect offsets, while the rest target defence specific or civil sector benefits'.[31]

On the other hand, at least some sections in the government are firmly in favour of *direct* offsets. The viewpoint from the government's perspective is that since the defence offsets have been introduced as an instrument to enhance defence production capability, the money influx through offsets should be absorbed only in defence production. 'The primary aim of the offset policy of the Ministry of Defence since its inception has been to utilize this massive investment in defence modernization to provide a fillip to the Indian defence industry both in the public and private sector. Though from time to time certain groups have voiced the view that the offset program in India needs to be relaxed to include other sectors, however, from the government's perspective it is very clear that offsets are primarily to strengthen the defence industrial base in the country'.[32] There are other issues too in part to avoid gaming on part of the OEMs and in part to ensure the other sectors are actually ready to absorb offsets: 'From a Finance viewpoint,

---

[30] Smita Purushottam, Chapter 6.
[31] S.N. Misra, Chapter 5.
[32] Shobhana Joshi, Chapter 7.

accountability concerns need highlighting. The primary purpose of offset policy should be increased indigenous production and access to critical technology. Indirect offsets should be secondary and need to be gradually rolled back as the industry's capacity to absorb direct offsets increase'.[33]

Enhancement of FDI

Analysts support enhancing FDI limits in defence offsets. 'FDI is the key to a business model and with an enhanced FDI, the foreign OEM has a sustained interest in the Indian partner and is likely to nurture the production to mutual benefit. As long as "mutual benefit" is the central theme, offsets will grow'.[34] Indeed, analysts recommend 'that FDI be increased to at least 50 per cent as this is likely to bring in significant, key manufacturing and design technology capability'.[35] The OEMs and offset recipients too are in favour.

However, the government, at least till 2014, disagreed: '... about 70 per cent of Defence equipment is imported. It seems quite anomalous that we should be willing to buy strategically important weapons and systems from foreign manufacturing companies, situated abroad, and not allow the same companies to invest more than 26 per cent in India and manufacture this equipment in our country. Surely national security would be better met if a foreign-owned company manufactured the equipment in India, compared to importing the same equipment. The logic that does not allow this to happen can perhaps only be explained in terms of a perceived threat to the DPSUs, the unions of the PSUs and the vote banks associated with these unions. There has been a considerable body of opinion in favour of enhancing this limit but so far the government has not made any change'.[36] The government that came in power in 2014 is amenable to relaxing the FDI limit.

---

[33] Vijayalakshmy K. Gupta, Chapter 3.
[34] K.V. Kuber, Chapter 8.
[35] S.N. Misra, Chapter 5.
[36] R.C. Bhargava, Chapter 20.

As may be expected, the results by way of foreign investment have not been encouraging till 2014 although there is renewed interest after the change in the government. 'While opening the defence industry to the private sector in May 2001, the Government allowed 26 per cent FDI. It was hoped that foreign investors would rush in with their bags of money. However, all hopes have been belied and the policy has been acknowledged as a total failure. Most prospective foreign investors view the policy to be highly dissuasive in intent and content. There has been a total lack of enthusiasm on the part of foreign investors'.[37]

Things began moving in 2014. The ruling party had promised a cap of 49 per cent across the board before coming to power in mid-2014, but as one OEM executive pointed out, 'an uplift from 26 to 49% maintains the status quo and may not be sufficient incentive to make an investment here'.[38] Indeed, in June 2014, the Department of Industrial Policy and Promotion (DIPP) circulated a discussion document to allow up to 100 per cent FDI in defence production in manufacturing of state-of-the-art equipment. The document additionally suggests a cap of 49 per cent for investments that do not involve transfer technology and a cap of 74 per cent where the foreign investor is ready to share technology know-how. The incoming government after the elections has pushed for 49 per cent, so we expect the policy to relax from the earlier limit of 26 per cent and possibly to 74 per cent and 100 per cent in some cases in 2015. However, as of this writing, this remains to be seen.

Capability Building

An effective way of capability building is through enhanced investment in research and development (R&D). DRDO is the predominant agency through which investment in R&D is carried

---

[37] Mrinal Suman, Chapter 4.

[38] Reuters, 29 June 2014. 'Eyes on defense deals, Western powers rush to court India's Modi', accessible at http://www.reuters.com/article/2014/06/29/us-india-defence-idUSKBN0F403P20140629.

out. The investments by the private sector are minimal with major companies like Larsen and Toubro (L&T) and Tata earmarking budgets for R&D in terms of decimal percentages of their revenues. 'With such scant priority to R&D, the capacity to absorb critical technology and manufacture major sub-systems would be grossly inadequate. The lesson learnt is that just by paying for technology transfer and entering into licence production, you are not going to have technology unless you have the R&D readiness and wherewithal in this country'.[39]

The unwillingness of private sector to invest in R&D is understandable from the analysts' viewpoint. 'Defence markets are not free markets, they are narrow, with oligopolistic and market imperfections'.[40] The major disincentive for investment is the risk involved in this oligopolistic market. With the government as the sole client for Indian industry and zero opportunity for exports (exports of weapons are not allowed as a matter of policy), the private industry is loath to invest in defence production. A change in government or a change in policy can sink an investment of thousands of crores of rupees (hundreds of millions of dollars). This disadvantage when combined with a slow gestation period is beyond the risk appetite of Indian industry. Hence, Indian entrepreneurs find it simpler to invest in R&D for producing low value items in the supply chains of global OEMs where the risk is also small. The government has to come forward to take initiatives in funding technology in both the private and public sectors. As one analyst notes, 'even today in the advanced world and in the most technologically advanced country, the US, the DARPA still funds the entire technological initiatives'.[41]

One possible way for the government to jumpstart R&D in defence sector is to encourage joint ventures (JV) in R&D between global OEMs and Indian industry. Green shoots became visible with the memorandum of understanding (MoU) between Sweden-based Saab and Mahindra Satyam for developing R&D and training

---

[39] S.N. Misra, Chapter 5.
[40] Vivek Rae, Speech at Swavalamban II Conference at Hyderabad, 2011.
[41] K.V. Kuber, Chapter 8.

centres and in another MoU between Saab and Pipavav Shipyards for technical collaboration in 2011. The defence offset policy needs to incorporate measures to promote such initiatives. However, such incentives are woefully lacking as noted by some analysts. 'Till date, despite more than US\$4 billion signed in offsets contracts alone, there is no single case for FDI in R&D; obviously there is something wrong with our formulation of the policy, and this has to be addressed by the MoD. This has not been adequately addressed in the offsets [policy of] 2012 either'.[42]

Some issues analysts point out are as follows:

- The scope of indirect offsets needs to be expanded. 'The offset policy in 2009 included civil aerospace sector in the ambit of offset. However, the commercial shipbuilding has been left out of the purview'.[43] Commercial shipbuilding, infrastructure and telecommunications are worthy contenders to be included in the ambit of defence offsets. These sectors require massive influx of capital and defence offsets holds promise of massive investments. If included, these would benefit many sectors.
- The process of intellectual property (IP) awareness management and IP process management are at a nascent stage in our country. There is a need for government support and understanding of IPR dynamics.
- The requirement of licensing has been an impediment for many companies who intend to get into manufacturing. 'This is essentially because of the fuzzy policies and lack of clarity'.[44] There is no clarity even on the issue that which products require a manufacturing licence and which do not. Most companies have a waiting period of 18 months and more even to just know if they require a manufacturing licence!

---

[42] K.V. Kuber, Chapter 8.
[43] S.N. Misra, Chapter 5.
[44] K.V. Kuber, Chapter 8.

- The present tax and duty structure treatment of offsets limits offsets to supply of parts and systems by Indian industry to OEMs by way of physical exports and thus misses out on system integration/manufacture within the country. Items outsourced for producing in Indian territory by OEMs, even if paid in foreign exchange are counted as offsets but not as deemed exports. The government needs to recognize the offset business as 'deemed export' to incentivize the OEM to develop a local vendor base.
- The tax structure appears to be loaded against domestic private sub-contractors. On one hand, sub-contractors for DPSUs are entitled from customs duty exemption and likewise excise duty exemption is being claimed in terms of Customs and Excise Duty Certification issued by DPSUs. The non-exemption to sub-contractors of the domestic private sector results in increase in costs. At the same time, input taxes continue to be added and loaded on to the bids for the domestic private sector as there are no exemptions to the sub-contractors. In contrast, foreign OEMs are able to minimize the input cost on account of export benefits available to them from their respective countries.
- The policy on defence exports is complicated by the stringent requirements of export control through SCOMET. The policy must be simplified to provide clarity as to which products and which countries are restricted or controlled. Also, in the current environment, companies have to wait 8–10 months to get an export clearance order.
- Establishment of aerospace and defence special economic zones (SEZs) has been for long under consideration by the government. Linking these with the NIMZ being established under the national manufacturing policy would synergize the operations of the Ministry of Commerce and MoD. Such parks would also provide a robust ecosystem for strategic manufacturing.

The Role of Defence Public Sector Undertakings

DPSUs have the lion's share of the domestic defence production in the current environment. However, not much progress has been made as reflected in the fact that India currently imports 70 per cent of its defence requirement. There is a need for change in methodologies and influx of modern technology to enhance this production qualitatively and quantitatively. 'What we are hoping is that the entry of the private sector would complement what the public sector is already doing. We wish the public sector industry to mature into integrators rather than being involved in manufacturing every small part themselves. We also hope that the capability of manufacturing sub-systems and competency in complex niche technologies will be taken over by the private sector'.[45] The DPSUs already have a manufacturing ecosystem with a large number of defence-related micro-small-and-medium-sized enterprises (MSME) built into their supply chain network. This base also needs to be strengthened by assisting the MSMEs in absorbing new technologies and providing them access to cheap funds to improve themselves.

Another role for DPSUs could be integrating with private sector to form public–private partnerships (PPP) with government assistance. One possibility is logistics. 'The [Armed] Services are reluctant to resort to outsourcing of tactical logistics as it forms the first and second line of support to troops and provides them confidence during operations. Outsourcing creates critical dependencies that do not bode well during crisis situations since legally, private contractors could not be compelled to go to war zones'.[46] A possible solution could be through forming JV companies and producing major systems with clear division of work and marketing the products with joint name of the DPSUs/Ordnance Factory Board (OFB) and the private sector JV partner.

---

[45] M.M. Pallam Raju, Speech at Swavalamban II Conference at ISB Hyderabad, 2011.

[46] Vijayalakshmy K. Gupta, Chapter 3.

Accountability of Offsets

A major issue during the selection of a successful bidder for defence contracts is that the offset offer is not considered at all while selecting the lowest bidder (L1) as the vendor. It is assumed that the offset component for all vendors would be similar and necessary amendments and incorporation can be done when the vendor has been shortlisted for the primary contract. 'Theoretically, the offset part of a quote should be treated at par with payment terms and be harmonized across all vendors before determining L1 because offset has a bearing on the quoted price'.[47] This is an anomaly that needs to be addressed at the earliest. Still, going by the press, this could have been a reason in India announcing the selection of the French company Dassault's Rafale aircraft in 2012 although even by mid-2014 that deal had yet to be inked.

This requires verifying in practice whether the contractual terms are being met. 'A strong audit mechanism is needed to enforce and verify compliance of offset obligations as false claims of compliance may go uncontested if there is collusion between foreign OEM and Indian offset partner'.[48] In the recent Defence Offset Guidelines, the MoD has attempted to strengthen the erstwhile DOFA by restructuring it with the hope that the new body (DFMW) would be able to carry out effective monitoring and audit of all offset deals.

PERSPECTIVES FROM GLOBAL OEMs

India is a growing defence market in an environment where the previous large buyers in Western countries are cutting down purchases. This makes India attractive to global OEMs. The OEMs believe that the relationship between the government and foreign OEMs is of paramount importance. 'The overarching requirement for a successful offset policy is a collaborative relationship between

---

[47] Vijayalakshmy K. Gupta, Chapter 3.
[48] Ibid.

the government customer and the foreign OEM in the successful execution of the offset obligation'.[49] As profit-making organizations, their goal is not social causes but to make money. 'The offset programme can be successful only if the environment for transacting business is conducive and all stakeholders can profit to make it a win–win situation'.[50]

Managers from OEMs who have contributed chapters to this book are ambivalent on the issue of balance between direct and indirect components in discharge of offsets. This is possibly because they as stakeholders are least affected by the direction this balance takes. The topics of interest to OEMs are (1) FDI in defence offsets, (2) institutional framework to manage offsets, (3) flexibility in implementation of offset policy and (4) potential models for the Indian private sector.

FDI in Defence Offsets

For the OEMs, the percentage of FDI in offsets is less about profits and more about a long-term association with the Indian industry and the government. However, the key issue that remains is whether raising the FDI limit to beyond 26 per cent and even beyond 49 per cent would bolster the Indian industry? 'What has been experienced in the course of several bids that we have made to the MoD is that FDI affects Indian industry in areas of technology, competitiveness and opportunities for export'.[51] Hence as per the OEMs, the government needs to ask itself whether taking the limit beyond 50 per cent would enable the domestic industry better in these three critical areas. 'OEMs would certainly welcome FDI going from 26 per cent to 49 per cent and beyond so that the expectations of the government, the industry and the OEM are met in a win–win manner that we all stand to gain out of this exercise'.[52]

---

[49] George B. White, Chapter 12.
[50] Nalin Jain, Chapter 10.
[51] Thelakat Jayadevan, Chapter 11.
[52] Ibid.

Yet, as noted earlier, other OEMs believe that going to 49 per cent does not really change anything in making India more attractive for investment.

The OEMs feel that the current limit for FDI is insufficient for them to put in the management efforts to bring in cutting-edge technology due to lack of control over the JV. They want to transfer technology but are looking for more incentives from the government. 'Raising the FDI threshold will increase the pace of advancement towards achieving self-reliance by Indian industry. It will encourage foreign companies to bring cutting-edge capabilities into India; to develop cutting-edge capabilities indigenously from scratch would be risky, time-consuming and expensive. Bringing capabilities on-shore can help transfer control and oversight to the government as compared to purchases made overseas'.[53] This is especially true in instances of critical times for national security.

Another aspect is of risk sharing. OEMs look globally for growth, profitability and for strategic alignments. 'As for strategic align-ments, aerospace is a very capital-intensive industry and to that extent, a very high-risk industry'.[54] The lower the FDI threshold, the more is the risk with possibilities of change in strategy. To the OEMs, a higher threshold would act as a signal from the govern-ment of reduced strategic risks.

Institutional Framework to Manage Offsets

The OEMs are wary of the lack of clarity in policy and perceive goal posts as continually shifting. Frequently changing policies and policy makers lead to mid-course revisions in interpretations where substantial financial investments have already been made, result in frustrations and sunk costs. As one executive put it, 'One of the biggest challenges that BAE Systems is facing and I am sure other

---

[53] Jeremy Charmak of BAE Systems in his talk on 'Challenges in graduating to be an offset partner in India', during seminar 'Swavalamban II' held on 8–9 December 2011 at Indian School of Business, Hyderabad.

[54] Nalin Jain, Chapter 10.

foreign defence companies are too, is the difference in interpretation and opinions and different approaches we take'.[55] As such, they say that 'care must be taken to ensure that consistent policy implementation continues in the face of ongoing personnel turnover'.[56]

The OEMs would prefer to achieve consistency through a centralized agency staffed with knowledgeable government officials who are not transferred out after limited tenures. 'There has been a lot of talk about a clear and consistent offset policy—to have a central management organization, which enunciates the policy, develops and owns the related regulations, and is accountable for implementing the arms and weapons acquisition process'.[57] From the OEM viewpoint, 'it is recommended that the MoD consider establishing a single offset authority as the most efficient means of ensuring consistent and effective application of the offset policy over the long term'.[58] The government in its recently released Defence Offset Guidelines (2012) has indeed taken stock of the issue and established the Defence Offsets Management Wing. However, the effectiveness of the new organization remains to be tested and continuity issues around turnover of government officials still need to be settled.

Flexibility in Implementation of Offset Policy

The OEMs would prefer that in the initial stages of any offset program only plans be submitted. They believe that there is a need for flexibility in having case-by-case implementation of policy rather than a blanket application of rules. 'A perfect set of offset rules and guidelines does not seem to exist as they need to be tailored to

---

[55] Jeremy Charmak of BAE Systems in his talk on 'Challenges in graduating to be an offset partner in India', during seminar 'Swavalamban II' held on 8–9 December 2011 at Indian School of Business, Hyderabad.

[56] George B. White, Chapter 12.

[57] Jeremy Charmak.

[58] George B. White, Chapter 12.

governments' policies.... The broadest and most flexible schemes with high incentives bear the most convincing results and hence are most successful in the long term'.[59] 'Flexibility means that business opportunities may be afforded to the local industry that could not even have been envisioned at the time the offset contract was defined. Flexibility also means that the government customer may revise their policy objectives during the course of an offset program, and still see those new objectives realized within that program'.[60]

To be fair to the Indian government, in a democracy, actions of persons in authority are subject to scrutiny and hence detailed justification needs to be incorporated when exceptions to policy are made. This becomes a challenge especially when the office bearers and decision makers have limited tenures in the positions they hold. The Right-to-Information Act permits scrutiny of old programmes and this enables any citizen to question the intentions of persons in authority at a later stage. As a result, decision makers practise restraint and do not undertake deviations in policy even though it may be beneficial to the overall growth of a programme or an industry. There is hence a need to set the policies in a format that has flexibility as an inherent component without having the need to resort to exceptions. 'In order to help realize win–win outcomes for all stakeholders, including the MoD, Indian industry, and the foreign OEMs, it is suggested that greater flexibility be accommodated by permitting changes in the offset contract with respect to projects, values and partners, with the written agreement of the MoD'.[61]

Potential Models for the Indian Private Sector

One of the key aspects of the Indian Offset Policy is that global OEMs are free to choose their Indian offset partner (IOP). This leads to the obvious question as to what do OEMs seek in these

---

[59] Yannis Ailianos, Chapter 13.
[60] George B. White, Chapter 12.
[61] Ibid.

partners before getting into a joint venture and what the potential business models for these Indian partners could be.

In the offset space, joint ventures are about ToT from the OEM to their Indian partner. 'Transfer of technology is about investment in people. It is not about putting people into various roles in an organization but about finding the right people, who can take more responsibility within the organization and develop it as it goes further. A lot of technology transfer that we talk about involves "know-how" and the "know-why"'.[62]

OEMs look for some specific aspects in their IOPs before engaging with them in JVs. 'We look at current capabilities of the IOP, we also look at the potential to develop new capabilities in the areas that we are looking for and in particular the understanding within the organization and the management and what are actually the steps required to take to achieve those changes. We also look at the culture of the organization … the management approach, how does the organization recruit, develop and retain people within the company…. We look at things like responsible business trading, corporate social responsibility and increasingly we are looking at organizations that develop sustainable operations'.[63]

There could be four potential models for OEMs to partner with Indian industry:[64]

1. A global sourcing model with Indian companies as Tier-1/Tier-2 suppliers
2. JVs with local partners
3. Global primes bring global suppliers to India and help them set shop
4. Risk-sharing programs with upfront investment by private companies.

---

[62] Jeremy Charmak of BAE Systems during seminar 'Swavalamban II' held on 8–9 December 2011 at Indian School of Business, Hyderabad.
[63] Ibid.
[64] Nalin Jain, Chapter 10.

Of these, the first two options are the ones discussed most in India. A large number of companies, including SMEs are trying to grow as defence and aerospace manufacturers. However, they can only succeed if their business plan is viable.

From the OEMs' viewpoint, offsets can enable Indian companies only if they have a commercial business plan and viable processes in place. Joint ventures cannot be based on subsidies. Eventually, offsets should assist rather than replace good business plans.

PERSPECTIVES FROM INDIAN INDUSTRY

Indian industry—both SMEs and the large companies—is the third stakeholder and important for both the government and the OEMs in meeting their respective objectives. Indeed, 'there is some element of entrepreneurship and efficiency that the private sector can add to the overall strategy in defence'.[65]

The long-term goal is clear: 'Ultimately, our aim should be to develop India as a centre of excellence for manufacturing regardless of offsets because it is the most competitive place to manufacture products'.[66] And Indian industry looks forward to graduating from 'build to print' to 'build to specifications'.

Offsets can be useful in the interim. 'Offset deals with Western nations provide a unique opportunity for the Indian manufacturing sector to grow and acquire competencies. We could either look at getting all the technology from outside—where foreign companies come and manufacture here—or we could pick specific areas for developing Indian competence'.[67]

Indian SMEs can play a major role. 'For developing India's defence industry, the incubation of the SME sector—which is of critical importance to the supply chain—is very important.

---

[65] V. Sumantran, Chapter 15.

[66] Vivek Lall, Chapter 19.

[67] S.K. Munjal, Foreword.

The SMEs are the pillars that can help in the creation of a true defence and aerospace industry in any country'.[68]

As far as the balance between direct and indirect offsets is concerned, the Indian industry as a stakeholder is firmly in favour of indirect offsets. One executive recommends, 'let us emulate the EU example and make it clear that 30 per cent is reserved for direct manufacturing in defence and total offsets will be much higher than 30 per cent and the balance above 30 per cent could be met through ToT, multipliers and counter-trade in other areas like civil aviation, homeland security and investments. Any policy change which would lower the percentage of direct offsets will be counterproductive to our primary National Objective'.[69]

Indian industrialists bring out four types of issues: (1) FDI in defence offsets, (2) exports policies and restrictions, (3) technology development and (4) SMEs as contractors to OEMs.

FDI in Defence Offsets

Like the analysts and the OEMs, the Indian private sector believes that the FDI limit of 26 per cent is a failure—even the 49 per cent mooted in 2014 is seen as challenging—and even the DIPP has proposed increasing the FDI to beyond the 50 per cent level in cases where technology is being shared. 'It seems that the restriction of 26 per cent FDI is hampering the inflow of technology from abroad. Those who have developed sophisticated technology are reluctant to part with this technology, unless they have control over the company getting the technology'.[70]

Enhancing the FDI limits will also bolster SMEs as the global OEMs would be more inclined towards joint ventures if they have some control over the company. Dual use technology obtained through such ventures could also be thereafter used in manufacturing other products not related to defence.

---

[68] Vivek Lall, Chapter 19.
[69] R.S. Bhatia, Chapter 14.
[70] R.C. Bhargava, Chapter 20.

Export Policies and Other Restrictions

The Indian private sector is disappointed with the arms export policy of the country. Arms exports, as an extension of national strategy, give tremendous political advantage to the exporter. Economically too, enhanced exports provide benefits of scale to domestic manufacturers and adds to the national GDP. However, one of the prerequisites to high-margin exports is that the exporting country should have access to cutting-edge technology not available with the importing nation.

In finding exports, Indian companies find all policies from licensing to production frustrating especially given the acute need for exports for the Indian economy at large. One industry executive notes: 'the export of defence products from India has been limited both by policy and the capability of meeting the needs of foreign countries. The extent of [annual] exports, [over 2007–2010], has averaged less than ₹300 crores [~US$50 million]. This limits the earning of scarce foreign currency, slows down the growth of manufacturing activity and creation of employment opportunities. The ability to increase India's influence in smaller developing countries in our neighbourhood is limited by our inability to sell required weapons to them'.[71]

Another executive refers back to history regarding the objectives of the offset policy when first conceived. 'When offsets were introduced as part of recommendations of the Kelkar Committee, the secondary consideration was to become part of the global supply chain thereby giving impetus to defence exports'.[72]

From the industry viewpoint, there is need to establish forward-looking export policy that takes into account benefits that can accrue out of defence offsets. With an industry-friendly export policy that eases licensing and expands the list of products that can be exported, companies would invest higher capital expenditure in defence production and take greater risks, giving them an upper

---

[71] R.C. Bhargava, Chapter 20.
[72] R.S. Bhatia, Chapter 14.

hand in negotiations with OEMs for technology transfers and joint ventures.

Technology Development

The value chain in technology development starts from *build-to-print*, then moves to *build-to-specifications* and finally results in *joint-technology development*. Indian industry is currently in the basic build-to-print stage and needs to do a lot in technology development to sign MoUs in joint research and technology development. Green shoots are visible with the recent MoU on technology development between Saab and Pipavav Shipyards, but there is a lot of ground yet to be covered. 'There are a few companies like L&T who actually invested a lot of money, built a lot of technology in different spaces and they have reached a level of competence where they are actually competing with a global OEM. But in the strategic electronics space [India] is still operating at the periphery and a lot needs to be done'.[73]

In most JVs that happen under defence offsets, the Indian partner does not have the necessary understanding of the technologies involved. 'The government has given them production licences to make it with the hope that they can go back and talk to a foreign vendor and eventually manufacture it. Foreign companies are not comfortable with sharing [intellectual property] and rightly so. They have spent years of R&D and have built this technology. They would be ready to share this technology only if they see value and find Indian companies at par in core technology development'.[74]

Moreover, Indian companies as offset partners do not have the technology readiness level to carry out core technology development. So here is Catch-22: the Indian partner wants a JV to gain technology from an OEM, while the OEMs will not partner till the latter has developed and demonstrated a minimum standard in core technologies!

---

[73] Arvind Lakshmikumar, Chapter 16.
[74] Ibid.

The logjam can be broken if the government takes steps to incentivize OEMs to do joint ventures even with partners that are not technologically enabled. This can be best achieved by giving control of the JV to the OEM. 'The government too is planning to open a window in the Offset policy whereby any buyback of R&D services by an OEM from Indian companies could be recognized as an eligible Offset. Investments by an OEM for R&D in Indian companies or for joint R&D with Indian companies in defence products could also be recognized for offsets'.[75] Indeed, after the national elections in summer 2014, the incoming government pushed for the FDI cap to be relaxed to 49 per cent, while as mentioned before, the DIPP has proposed raising this to 100 per cent, 74 per cent and 49 per cent depending on the criticality of the need and whether or not technology was being transferred.

SMEs as Prime Contractors

Indian SMEs would like to position themselves to deal directly with the Indian government as end customer on some deals. 'The prime contractor interacts with the customer directly. We have gone through this process and we feel that there is a lot of value [that] can be captured if you are the prime contractor. One virtue that will be helpful while dealing with this segment is: Patience. Patience will see you through difficult times. It may be three or four years before you get any order but you have to sustain. It helps to be low cost and efficient. Unless you are low cost and very efficient you will not be able to sustain the lean periods'.[76]

There is recognition and appreciation that 'the government of India has taken steps for MSMEs to be prime contractors or preferred partners to OEMs in joint ventures. The MSMEs till recently did not find a place in the Defence Procurement Procedures (DPP). The first sign of inclusion was in the form of a multiplier accorded to the OEMs for choice of an IOP in their offset proposals, all this

---

[75] Vivek Rae, Speech at Swavalamban II Conference at Hyderabad, 2011.

[76] Ashok Atluri, Chapter 18.

is still voluntary. 'There is a need for the DPP to address the MSME segment more poignantly and make a deliberate inclusion in the main procurement procedures itself, both for Capital and Revenue acquisitions'.[77]

Moreover, SMEs believe they are an option for prime contractors. 'The "make" procedure that we have under the DPP addresses the issue of indigenous design and development and generates competition among Indian companies for this purpose. They get into the competition, there is a down selection, two companies get shortlisted, they prepare DPRs, they prepare prototypes, submit bids and then assured orders are given to the lowest bidder'.[78] Managers from SMEs recognize that 'the L1 process [selecting the lowest bid] is the biggest benefit to the Indian SMEs. Once you qualify technically and your financial bids are open what they do is they compare you with the other bidders and if you are L1, they place the order with you as long as it is within the budget'.[79]

AS A RESULT ...

Based on feedback from stakeholders on the effectiveness of various drivers, the government has been evolving the DPP over the years. Major changes in the DPP ever since 2006 are shown in Table 1.3.

A number of industry demands have been incorporated in the DPP. There is, however, key aspects like increasing the FDI limit in defence offsets and the evolvement of a National Offset Policy that still remain. It is hoped that stakeholder feedback would keep reaching policy makers and help the government evolve a viable and practical defence offset policy.

---

[77] K.V. Kuber, Chapter 8.

[78] Vivek Rae, Swavalamban Conference, Hyderabad, 2011.

[79] Ashok Atluri, Chapter 18.

Table 1.3 Evolution of Defence Procurement Policy in India

DPP	Key elements
2006	<ul><li>Only Indian **defence** industries permitted</li><li>Defence products and services **not** listed</li><li>Offset credit balance **not** permitted</li><li>Multipliers and ToT **not** permitted</li><li>**No** access to military's 15-year long-term integrated perspective plan</li></ul>
2008	<ul><li>Only Indian **defence** industries permitted</li><li>Defence products and services listed</li><li>Offset credit banking **not permitted**</li><li>Multipliers and ToT **not** permitted</li><li>**No** access to military's 15-year long-term integrated perspective plan</li></ul>
2011	<ul><li>Indian industries with preference to DPSUs</li><li>Eligible products namely civil aerospace and homeland security products and services permitted</li><li>Offsets credit banking permitted</li><li>Multipliers and ToT **not** permitted</li><li>**No** access to military's 15-year long-term integrated perspective plan</li></ul>
2013	<ul><li>Indian industries with even playing fields to private industry and DPSUs</li><li>Eligible products namely civil aerospace and homeland security products and services permitted</li><li>Offset credit banking enhanced to 7 years</li><li>Multipliers and ToT permitted as part of offsets</li><li>Access to military's 15-year long-term integrated perspective plan</li></ul>
2015? (proposal in circulation)	<ul><li>Increase FDI limit to 100 per cent for critical technologies</li><li>Increased limit to 74 per cent where there is technology transfer (and keeping it at 49 per cent where this is not the case)</li><li>Establish a fund for enabling R&D in defence production</li></ul>

PART II
VIEWS FROM THE INDIAN
GOVERNMENT AND ANALYSTS

2

India's Needs and Defence Offsets

E.S.L. Narasimhan

Abstract

The author provides a holistic context of defence offsets vis-à-vis national security in this article. Though the domestic Indian companies have moved up the value chain, they have not been given their due share. Technology infusion is critical for taking defence production to the next level. The need of the hour is creation of a new breed of joint ventures and special purpose vehicles. Industry and defence labs should have greater, secured and controlled interface for cross-pollination of ideas and to facilitate transfer of ideas from labs to shop floor. At the end of the day, the weapon in the hand of the soldier should be there by virtue of being the best and to have the capability to defend our borders.

INTRODUCTION

It is increasingly recognized by the world community that India would continue to play an important and appropriate role globally. India has grown in stature among the comity of nations, being the largest democracy, as an emerging economic superpower and a nuclear power to be reckoned with. We must also recognize that today we live in a very complex security environment. This is not a choice that we have made for ourselves but we have inherited it. We per force have to be watchful and vigilant, secure our borders and be in a high state of preparedness. We are battling cross-border terrorism as well as several insurgent groups whose agenda is inimical to our national interests. Our defence capabilities have to rise up to the challenge and demands of the security environment over the next few years.

When we talk of defence, it is presumed deterrents are the best form of defence. We are really looking at manufacturing or making equipment where the deterrent effect would be much more than defence. If one has a good deterrent system, it is better than any form of defence. But then the question arises as to how we really build up our deterrents' capabilities. From 1947 to 1991, defence projection was in the public sector; thereafter it moved into the private sector; participation moving on gradually with the foreign direct investment being allowed into these industries.

The promulgation of the Defence Procurement Procedure (DPP) 2006[1] provided a thrust for enhancing the role of the private sector in defence protection. The Ordnance Factories and Defence Public Sector Undertakings have their own limitations to assume a high level of self-reliance. With more and more development systems getting user acceptance, the capability to build complexity in the system and platforms have been established in the country.

However, these are not getting into production in adequate numbers. The existing production capacities and because of that the self-reliance index has not seen much of a change. Now there is increased realization that the goal of achieving self-reliance in indigenous development of defence systems cannot be achieved without enhancing the defence industrial base in the country and that would enhance production of indigenously developed systems. The only way to achieve this is by creating additional capacities in production, in both public and private sector.

India is currently at an inflection point vis-à-vis the choices it has to make to build up the defence capabilities. Self-reliance as well as empowering of nascent defence industry is an important

---

[1] Editors' note: DPP 2013 has since been released and a discussion paper, possibly for DPP 2015, is in circulation for comment.

recommendation made by the Kelkar Committee. India's defence production policy aims to enhance the capacity in defence systems and equipment. The development of missile systems and other weapons call for establishing indigenously advanced technologies in various areas and we have succeeded in that. Defence Research & Development Organisation (DRDO) has involved both small- and medium-sized enterprises and large private sector industries in a very big way in development of various systems.

Having been involved in various development cycles, many of the Indian private sector industries have today moved up the value chain from build to print, build to specification, build to design and build to requirements. Some of these companies can also take up the responsibility of lead system integrators and many of them can manufacture technologically advanced products that are world-class. There are about 500 Indian private sector companies all over India, which are capable of manufacturing high technology systems, sub-systems, components in the fields of design, avionics, limited effect weapons (LEWs), etc. This has catapulted many industries from supplier of products to DRDO to supplier of products to multinationals. Thus, the capability established in these firms is exploited by one and all.

Unfortunately, somehow in India we really have not given these companies the due share they require. Technology infusion is imperative and must be made integral in methods of discharge in the offsets policy.

TECHNOLOGY UPDATES AND UPGRADES

When one talks about technology development/infusion, there is a need to distinguish between upgrades and updates. The thin line of difference between these two is difficult to comprehend. As a result, much of our equipment is probably far behind times. Given the current trend of defence expenditure, expert opinion indicates that India's imports in the decades beginning 2011 could be US$100 billion. At the current rate of offsets, the opportunities available to the domestic scale enterprise could be in the range of US$30 billion. These are enormous opportunities waiting to be harnessed as well as challenges that need to be overcome. Indian industry has not gained in terms of capability creation. Other than contract manufacturing, which includes print to manufacture, there has been little evidence of capability building as such.

The Indian defence industry needs technology for growth and not just mere business in terms of contract manufacturing. Self-reliance cannot be achieved by the industry if it remains a sub-contractor to foreign original equipment manufacturers (OEMs) or content with contract manufacturing areas determined by the OEMs. It is widely believed that the offsets have a cost that a buyer pays and the seller caters for in his bid. There are two distinct elements in cost. One is that of discharging offsets and the other is the administrative cost of offsets. We need to see how we can narrow down these costs so that the prices can be kept under control. It is therefore imperative that the risks are mitigated and reduced by the OEMs to bring down the cost of acquisition. Infusion of technology-related aspects as one of the means of discharging offsets will further help in expanding the scope of discharge mechanism and thus help in reducing the cost of the main acquisition programme.

ACQUIRING TECHNOLOGY

In the last three years, a serious and concerted effort has been made by the government to reform and streamline the entire acquisition process. The government has come to appreciate the potential of the private sector and wants it to complement the efforts of the public sector. A number of initiatives have been taken and policies reviewed. *Yet the results on the ground have not been up to pace.* There has not been an appreciable inflow of anticipated foreign funds. Even today, all the major defence deals are signed with foreign OEMs. The public sector continues to get bulk orders under transfer of imported technology. The private sector continues to be a peripheral participant with the production of some low-tech items. The present process of interaction and integration should be continued albeit with renewed vigour and purpose. All joint committees should be represented at the level of decision makers so that the follow-up action can be implemented in a time-bound manner. Technological progress of the private sector should be given due recognition and considered a national asset.

The objective of achieving self-reliance remains elusive unless the private sector is duly integrated and its potential fully harnessed to build a viable indigenous defence industrial space. The government has to create an environment where the private sector feels assured of just business opportunities, level playing field and fair play. However, some critical issues have to be addressed to infuse a greater level of confidence as well as certainty. The ultimate objective is to create an ecosystem with greater institutional support mechanisms. This would call for a more proactive approach overall. Offsets were introduced with the sole aim of strengthening the Indian defence industry. Capability building cannot take place by contract manufacturers alone or by outsourcing work in areas that are non-critical to the sector.

Technology is the key to manufacturing high-end defence products in the medium and long term. This aspect alone has the ability to transform capabilities in the domestic industry. If we desire to reverse the indigenous to import ratio of 70:30, we have no option but to focus on technology acquisition. The experience thus far has proved that technology cannot be bought off the shelf and such a buy will invariably be riddled with restrictions in terms of licensed manufacture with little scope for modifications or upgrades to the technology offered. Vital elements of technology transfer reside in its absorption and harvest to propel future systems development. In a competitive scenario, discharge of offset obligations should trigger the OEMs to consider the projects with high-end technology infusion with global manufacturing entities being established in India. The key challenges are there in the domain of advanced technology products and prototype development; and these would call for special support and funds.

DEFENCE TECHNOLOGY DEVELOPMENT FUND

Most countries in the forefront of technologies have several in-built mechanisms and budgetary provisions to fund and support their industries to take up the challenging R&D work for military applications through innovative ideas. Recognizing this factor, the Kelkar Committee had recommended creation of a defence technology development fund for use by the Indian private sector industries to undertake high-risk development task. It is essential that this corpus fund is created without any further delay. Creation of testing facilities and the development of advanced manufacturing technology hubs require special interventions by leveraging on PSU resources and creating infrastructure of entirely different magnitude with wholly different boundary conditions. We should foresee and facilitate creation of a new breed of joint ventures and special purpose vehicles. Industry and defence labs should have greater, secured and controlled interface for cross-pollination of ideas and to facilitate transfer of ideas from labs to shop floor.

At the end of the day, the weapon in the hand of the soldier should be there by virtue of being the best and to have the capability to defend our borders. We cannot compromise our national security because of the criterion such as preferred vendors. We will have a dependable and secure supply chain; this would empower our security instead of depending on tenuous suppliers. Overall, the security frame of needs should be re-jigged in the context of a greater role for private enterprise with adherence to an all-encompassing security protocol. In the balance, defence offsets present a challenge of a new order and dimension. There is a rewarding opportunity on the anvil and we need to manage the situation by addressing the complexities inherent.

TAKING A HOLISTIC VIEW OF NATIONAL SECURITY

Today national security has undergone a paradigm change and it is no longer just defending borders, it is no longer restricted to military warfare or fighting between armies. Today nations prefer a low-cost war with lower human casualties and less financial implications. That being the case and with national security concept having undergone a paradigm change and bringing in concepts like food security, health security, energy security and change in the internal security scenario and the cross-border defence scenario, private sector has a major role to play. This is particularly because technology has taken over ideology after the crackdown of the erstwhile Soviet Union. India, however, is not really focusing greatly on technology, but on contracts, money machines and so on. The focus is also divided between health security, energy security and food security amongst others. There is an urgent need to overcome this attitude of compartmentalization.

Consolidation is an important component of national security; the lack of which is causing sufferings to our nation. Today one talks of massive industrialization and various memorandums of understanding (MoUs) being signed for various projects resulting in agricultural land being taken away to set up these projects. With lesser land to control how does one improve or enhance the agricultural output to ensure food security? This is a paradox for the planners with population that is growing on the one hand and agricultural land that is shrinking on the other. We have to strike a balance between industrialization and agricultural land. How do we use newer technology to ensure better agricultural production, to ensure good security, and a healthy India? Unless we have a healthy nation and healthy individuals, where is the workforce going to come from? We cannot import workforce. We have a billion plus hands to feed in this country. How are we going to ensure that this population is a healthy population that can go to work on a regular basis and produce things? Energy security is another topic of concern and how far are we going to ensure an assured supply of fuel to meet our energy needs and give us energy security? Today, a country is required to be an economic power rather than a military power. If one wishes to be a superpower, economic power is more important than military power. To be an economic superpower you need to look at food security, health security and internal security.

One needs consolidation of all forces for national security. Are we really focusing on these issues? The private sector really has a role to play in each of these aspects. They can contribute by way of technology and concepts. The government alone cannot do all these things however brilliant it thinks it is.

THE NEW THREAT OF CYBER WARFARE

Cyber warfare is a major current threat. How far are we equipped to fight this threat in the country today? We are networking everywhere. We are saying that we are going ahead around the world where globalization has come in. We have got everything wired up and with the click of a button one can get all the data. But does one realize that the enemy also can get all the data by the click of a button? How far are we protecting ourselves from that? That is where the private sector can contribute in a substantial manner. They can get various technologies from all over the world whereas the government has to face restrictions everywhere.

CONCLUSION

Defence of a country cannot be carried out by weapons alone. Each department cannot live in its own domain. Until the entire economy prospers, until our country becomes an economic superpower, self-reliance or *Swavalamban* will be difficult to achieve.

3

Offsets: A Financial Perspective

Vijayalakshmy K. Gupta

Abstract

In the post-liberalization era, Indian industry has significantly developed its capabilities and competitiveness but its R&D content is limited. It is more about licensed production and contract manufacturing. Transition from the world's back office and bulk manufacturing to the front end development and precision manufacturing is the need of the hour for our country. In this background, the offset policy of defence acquisition should be so leveraged so as to build a national industrial capability for self-reliance in defence preparedness. There are several issues, such as the capacity and capability of Indian industry to absorb the offsets especially on the technology side, costing of offsets especially indirect offsets, changes in offset portfolio at post-contract stage, desirable offsets vis-à-vis offered offsets and our desire to seek system integration in India, which need to be analysed and then tackled. The primary purpose of offset policy should be increased indigenous production and access to critical technology. Indirect offsets should be secondary and need to be gradually rolled back as the industry's capacity to absorb direct offsets increases. From finance viewpoint, accountability concerns must be taken note of and addressed with the seriousness they deserve.

BACKGROUND

Colonial rule had ensured that India missed the Industrial Revolution and our country did not have a strong industrial base when we became independent. Our resource-strapped nation has to catch up with huge arrears in building, upgrading our infrastructure and meeting social needs. This dilemma between social/ physical infrastructure and defence continues to vex policy planners at different levels. As a developing country exposed to various internal and external security threats, we cannot afford to let our guard down. We need innovative solutions to augment our defence capability as well as our indigenous defence development and production capability in pursuit of the goal of self-reliance. We are a young nation with more than half of our population being under 25 years. We must reap this demographic dividend by augmenting our manufacturing capabilities. In this broader context, getting appropriate returns for our demand in infrastructure and defence equipment through increased growth of the manufacturing sector assumes great importance.

The offset policy being pursued by the Ministry of Defence (MoD) reflects our desire to leverage our big-ticket defence acquisitions to create jobs and capabilities at home. Under the offset norms, foreign companies bagging government contracts are required to invest a portion of their earnings back as investments into India. Building national industrial capability for self-reliance in defence preparedness is our goal and we have to see how best to achieve this within the complexities of international market for defence hardware.

Close combat will always remain but future wars will certainly involve more and more push button technologies as deciding factors of the outcomes. Therefore, we need capabilities in areas of electronic warfare, intelligence gathering and analysis, tamper-proof communications systems and overall cyber security of increasingly computerized national wealth.

INDUSTRIAL CAPABILITY IN INDIA

In the pre-liberalization era, despite greater emphasis on industrial growth in the 2nd Five Year Plan onwards, the industrial activity was allowed only under licence, imports were controlled and a cap was put on the production capabilities. This had affected capacity building and technological advances across sectors. The post-liberalization era saw the removal of import restrictions, thus bringing in competition from the global players. Indian industry thus has moved slowly but steadily towards building a strong industrial base with a successful track record of implementing technology-intensive projects, including bulk production, within stipulated time frames at reasonable cost and of world-class quality. It has its strengths in design, engineering, finance and marketing. It has a reservoir of management, scientific and technological skills. The growth in the manufacturing sector has been significant and global standards as a norm have been achieved in engineering and manufacturing. Manufacturing now accounts for about 14 per cent of India's GDP, 35 per cent of foreign direct investment (FDI) and employs 12 per cent of the workforce. India's competitive advantages offer huge opportunities for increasing the exports.

According to a study by the Boston Consulting Group, India's vast domestic market and relatively low-cost workers with advanced technical skills will make it a manufacturing powerhouse within the next decade. Therefore, multinationals have already started setting up operations in India to operate in skill-intensive industry segments requiring advanced technical expertise areas in which India is becoming a primary sourcing and manufacturing base. In fact, high skill sectors account for sizeable percentage of the manufacturing output of India. However, it is also to be noted that most Indian industries, whether public or private, have been engaged in licensed production of foreign technologies and their own R&D contribution has been minimal.

Easy access to Soviet platforms and weapons ensured that defence-based industries did not come up in a big way. Our policy of licence, permit, quota and non-export of arms and ammunitions was a dampener and this policy saw to it that private sector had neither incentive nor encouragement to invest in this field. With the collapse of Soviet Union, we had to change our strategy. Massive indigenization was necessitated for repair and maintenance of the Soviet origin defence equipment. The liberalization of the Indian economy that almost coincided with the collapse of Soviet Union unlocked India's productive capabilities and gave impetus to the goal of developing a domestic military–industrial complex. In defence sector, there are large and small industry houses that have, over the years, built capabilities and capacities, through partnership with development agencies such as Defence Research & Development Organisation and production agencies such as Ordnance Factories (OFs) and Defence Public Sector Undertakings (DPSUs).

Department of Defence Production has looked at the issue of enhancing indigenization in defence, by increasing Indian private sector participation. Offsets to defence acquisitions provide an excellent opportunity to further develop defence industrial complex with private sector participation. Increased outsourcing is generally recommended as a prescription for enhancing the defence industrial capability. The DPSUs and OFs are increasingly becoming 'system integrators' and outsourcing sub-assemblies/sub-systems to private sector. As the capabilities and confidence develop, more and more critical sub-assemblies may be outsourced although it can be seen that outsourcing has its own limits. Beyond a certain point, it becomes financially imprudent to do so due to unrecoverable fixed overheads.

It is natural that Indian companies are not content with being mere suppliers of the DPSUs/Ordnance Factory Board (OFB). They aspire to be system integrators themselves or want at least some sort of co-branding. In a traditional outsourcing model, the contribution of the 'outsourced' supplier does not get adequately recognized although he may resort to some sort of branding, if the supplied parts are visible, e.g. a tyre supplier to a car manufacturing company can retain his visibility by putting his brand name/logo on the tyres. Such visible branding of components/sub-systems is not always feasible in compact defence systems. To address the problem, the desirable route is to form joint venture companies and produce major systems with a clear division of work and marketing the products with joint name of the DPSUs/OFB and the private sector JV partner.

One of the oft-quoted reasons against outsourcing in defence is the lack of confidence in the ability of the private sector to meet the requirement during times of crisis. It is understandable that the services want to tread the outsourcing path cautiously. The services are reluctant to resort to outsourcing of tactical logistics as it forms the first and second line of support to troops and provides them confidence during operations. Outsourcing creates critical dependencies that do not bode well during crisis situations, since legally private contractors could not be compelled to go to war zones. It could be understood as to why the core functions like first line maintenance may not to be outsourced. However, there could be areas that need to be identified for outsourcing from private sector participation would mean greater indigenization and self-reliance in defence. Indian industry is being encouraged to enter into collaborations/partnerships with their foreign counterparts and thereby ensure supply of latest technology to the different arms of the defence forces. But there are practical problems in such an arrangement, viz. in case of obsolescence of products, the outsourced agencies find it commercially unviable because of the changing situations with respect to design, development and cost escalation over a long period, resulting in low availability of product upgrades. Besides, a limited vendor base escalates repair costs. It has also been seen that greater concentration on high-end products by original equipment manufacturers (OEMs) creates problems for product upgrades.

Earlier, the Indian industry was not sufficiently well developed and the defence services grew into an inward looking empire where virtually all their requirements were to be met through in-house arrangements. Military (dairy) farm and clothing factories are two such examples. Today Indian industry has considerably developed in various sectors and is meeting the requirements all over the country. Therefore, the defence services do not need to maintain huge transport fleets and depots to stock rations and other consumer items. After the success of Operation Flood I & II, they do not need to run military farms except in remote border areas. The continuous resistance to outsourcing has adversely affected the teeth to tail ratio in defence services. There is a need to remove this mistrust of industry through appropriate bold administrative decisions and if need be, through legislative means. The Defence of India Act and emergency ordinances already provide for emergency powers to the government that enables it to requisition all assets/services in the eventuality of war, but this may be supplemented with separate legislation on regulation and development of defence industries. The legislation may provide framework for registration/licensing of suppliers to defence services or DPSUs/OFB with obligations and special concessions in recognition of their contribution to country's defence preparedness.

The process of integrating the private sector in the defence industry in a substantial manner was initiated by the government in 2001. The Defence Procurement Procedure 2002 (DPP 2002) turned out to be the watershed for the defence industry as it allowed in-principle the participation of the private industry in defence production. The Kelkar Committee was constituted in 2004 to review private sector participation in defence production. Some of the recommendations made by the committee have also been implemented through the DPPs. Recently the government has notified the liberalized foreign investment regime in the defence sector by raising the 26 per cent limit to 49 per cent through approval route. Above 49 per cent the proposal will go to Cabinet Committee on Security (CCS) on case-to-case basis wherever it is likely to result in access to modern and state-of-the-art technology in the country.

The offset policy aims to develop the local defence industry. Investment in small and medium enterprises (SMEs) will carry extra points in the new offset regime. SMEs would be given a 'multiplier' of 1.5, which means an offset investment of ₹100 crore will be counted as ₹150 crore.

Defence industry is not restricted to a particular sector, and spreads across complete range of goods and products involving

Land Systems, Aviation, Marine, Arms and Ammunition, IT and Communication, Missile, General Stores, to name a few. There will be a need to define the priority areas and match them with the existing and potential capability of Indian defence industry for achieving optimum benefits by joint ventures. In the past, little Transfer of Technology (ToT) actually happened from foreign technology sources in the cutting-edge technology areas to the OFs and DPSUs. These organizations did master the production skill sets, however, the ToT model for production denied development of upgrades and new systems.

LEVERAGING TECHNOLOGY IN DEFENCE

The defence systems involve three distinct categories of technologies. One is about the chemistry of explosives of different varieties and another about the physics and the mechanical engineering of the various tanks, guns, rockets and missiles to launch the explosives to aim at the target. The third category involves high-tech electronics and computer engineering to develop various types of sensors, communication systems, guide the launch mechanisms and in fact the entire war machineries. The brute force and the destructive power of these explosives have to be guided to minimize collateral damage, which means precision strike technologies requiring sensors to guide the missiles. Thus, related are the high-end missile sensors, radars, aerostats, satellites, electronic warfare systems and communication networks and complex avionics. It is in this stream of technologies where India holds a great future. We have to master and improve these technologies, use all the bargaining power that our large defence spending programmes provide us to establish and scale up our capabilities.

We cannot remain content with getting manufacturing share in only the low-end, low-value bulk manufacturing containers, launchers, ground systems and vehicles. A number of parts or components in these segments involving high precision and high reliability manufacturing technologies are still imported. Besides up-scaling our production capacities in these areas, we also have to move further up in the value up-scaling of our conventional production capacities and enhance the value-addition by manufacturing high value, complex sub-assemblies of defence systems. These involve mastering several critical technologies. There is a need for India to increasingly focus on cutting-edge technologies research and product development rather than offering only back office and software program services. We need to transition from the world's back office and bulk manufacturing to the front end development and precision manufacturing. Investments worth more than ₹55,000 crore have reportedly been proposed by Indian and multinational firms to set up semi-conductors and other types of high-tech units in the country. The focus of offsets has so far been primarily on the defence sector though aerospace has also been added to address concerns about the Indian defence industry's capability to absorb offsets. Perhaps in the future, we may also expect further broadening of the scope to other areas such as Information Communication Technology that is extensively used in both civilian and defence sectors. Indigenous development and access to offsets to high-tech dual use technology should be pursued with utmost rigor and expanding the scope of offsets should be considered in this context.

COST AND BENEFITS OF DEFENCE OFFSETS

Self-reliance in defence is the stated policy goal of the government. This is the ultimate insurance against any form of denial of technology or capability. Not long ago, defence and its economics were considered a secret pursuit excluded from the public scrutiny. But this perception is fast changing. Defence is not a dead weight on the economy. The expenditure on defence provides the nation a secure environment for progress and it contributes to the industrial development, technology development and job creation. Economic benefits of defence expenditure can bear further analysis, quantification and awareness.

As succinctly put by Stephen Martin in his scholarly essay on 'Overview of Theory and Evidence (on offsets)', the following economic motives drive governments to seek offsets:

> Market Oligopolies, second best outcomes and capturing economic rent because the markets for high technology defence and aerospace products are characterized by oligopoly. Prices do not automatically adjust to changing market conditions. In such markets, purchasing governments may seek to use industrial policy in the form of offset requirements to achieve what the economists call the 'second best outcome' and to secure for its own citizens some of the economic rent that would otherwise have gone to the selling party. Economic rent is the extra payment being generated for a factor of production or an input, over and above what it would have earned in its most optimal use.

In competitive markets, economic rents are supposed to be just 'competed away' by new entry of firms. But in an imperfect market, such as an oligopoly, factors such as a steep learning curve, or need for high levels of investment, etc. may curb the entry of new firms and cause economic rents to persist. Offsets strategy may be used to compensate a part of this economic rent.

Technology transfers and spill-overs are another driving factor for offsets. These are the means of infusing new technology into the economy, thereby ultimately stimulating economic growth. The argument is that the social benefits of technology induction exceed the individual benefit to the firm or industrial enterprise. Thus, the firm will purchase new technology only up to the point where its marginal benefit equals its marginal cost. Hence, the government may use industrial policy (offsets) as a way to induce higher levels of investment in technology with spin-off effects for the entire economy. Are offsets the best method of acquiring the technology? Here the argument given is that they are more efficient than a straightforward technology transfer. The risks in a straightforward technology purchase are on the buyer, whereas, when the technology transfer is part of a larger contract, the reputation of the vendor is also at stake for the entire system, if he fails to transfer the technology successfully.

Protection to infant industry is yet another reason for offsets. Every government wants to increase its military capabilities by strengthening the country's defence industrial base. Where there is a steep learning curve and large investments, a certain level of sales is absolutely essential in order to bring down unit costs of production to global levels. This can be achieved through an offset strategy.

Overcoming barriers to market entry and protectionism can also be achieved through offsets, which can help the sub-contractors of the purchasing country to get into the bidders list of the manufacturers in the selling country. If the selling country has a large domestic pool of sub-contractors, or other kinds of barriers to entry, this may otherwise not have happened. Offsets in this case help to assist in the entry of foreign firms into a protected market.

Employment generation and regional balance also guide offsets. As in the case of technology, the gap between social cost and benefit and private cost and benefit is also used to justify the application of offset policy to generate employment of labour and to correct regional imbalances in development. An employer will recruit persons up to the point where the marginal (private) benefit is equal to the wage rate, but there are social benefits of removing unemployment, which may go unaccounted. Government, through the offset policy, intervenes to encourage firms to employ more people than they would have from a purely private point of view. In the same way, firms can be encouraged to relocate activities in specific geographic areas.

Most governments leverage purchases of defence equipment through some form of offsets. There are no universally laid down parameters or measures to weigh costs and benefits of offset programs. Even if some countries have individually undertaken exercises to evaluate such costs and benefits, the information, particularly the quantitative data, is not always available in the public domain. Studies have been conducted about whether offsets mitigate or magnify the military burden. Offsets have been termed as 'smoke and mirrors', with no conclusive evidence to establish economic efficiency of offset transactions to the buyer. For instance, in a survey conducted in the UK, it was concluded that 'evidence suggests that offsets do cost more than off-the shelf purchase and, not surprisingly, that vendors seek to include most of this premium in the selling price'.

Defence acquisitions involve huge expenditure and impinge on international relations. Arguments regarding the technological advantages and the trade and employment creation benefits of offsets often help in taking decisions about large defence purchases that otherwise may not have been possible. Other arguments that are often made are regarding security of supply, lower 'long run' or lifecycle costs as compared with the provision of services by the original manufacturer. In the last decade, global defence industry has gone through a lot of churning. So many mergers and amalgamations have taken place. Transnational alliances have been forged or strengthened. The companies have re-organized their production plans to focus on their core strengths and develop value-addition chains running at global scales. Big companies focus more on system integration and quality control rather than trying to produce everything in-house. DPSUs also need to follow this model and contribute to the development of military industrial components. Naturally, when OEMs of weapons, systems and platforms are asked to meet the offset obligations they cannot do so without the consent and cooperation of their sub-vendors who often hold intellectual property rights of these systems. Realizing this, the Government of India, therefore, has changed the offset guidelines to allow sub-vendors to discharge offset obligations.

ONGOING ISSUES WITH OFFSETS

From a theoretical perspective, there are four compelling factors—cost minimization/value maximization, resource access, superior resource leverage and risk diversification—that drive an organization to opt for outsourcing. The outsourcing decision is based on 'core' and 'non-core' activities that determine which activities need to be retained in-house and which are to be purchased from the market. However, in the extremely competitive global market, there exists a very thin demarcation line defining 'core' and 'non-core' activities. Also, the outsourcing decision taken by an organization is based on estimation of economic costs and benefits and on the assumption that transactions are carried out to the optimum ability of the organization in drawing and enforcing proper contracts—the last being susceptible to ignorance of market-player behaviour, thus becoming the key determining element for the success of any decision.

From the perspective of industry, the Indian private industry is suitably placed to work closely with their defence public sector counterparts. Private industry capabilities have not been optimally harnessed for domestic defence requirements, while in the global front, the industry has been successful in providing services to large multinational companies. While the benefits of defence outsourcing to the private sector is endless and includes reduction of import dependence, technological self-reliance, faster and cheaper supply of logistics, the areas where they can play a meaningful role is extensive and far reaching. However, not so long ago, the private sector's role in defence was confined merely to providing raw materials, semi-finished products and parts and components to OFs and DPSUs. But, with the government allowing private participation including FDI, the private sector has been able to produce high-end defence products, demonstrating its effectiveness in translating its civil prowess into defence business. With the rising profile of Indian private industries on the global scene backed by competitiveness, quality standards and efficiency there is enough confidence among the private enterprises to delve into hardcore development and production of hi-tech defence equipment. To accommodate private industry, in recent years the government has initiated policy measures to boost greater private participation. But what is needed is that along with policy measures, there should be a concerted effort to encourage public–private partnership (PPP) and greater outsourcing to the private sector. Some of the items that could be easily outsourced to the private sector include land systems, vehicles, engineering equipment, marine systems, aviation systems, arms and ammunitions, unmanned aerial vehicles (UAVs), surveillance systems, communication radars, radio sets and IT. Besides, the existing Defence Supply and Storage Network, vast Defence Estates, Armed Forces Medical Services and Defence Personnel Recruitment could be outsourced to private players for better management, speedy and quality services and for revenue generation. Similarly, the government should address the concern of the private players arising out of the discriminatory benefits enjoyed by the public sector in producing equipment, the prototypes of which are developed by private players.

There are a number of issues peculiar to offsets that have been come across. In a deal like acquisition of MMRCA wherein offsets have been stipulated at a whopping amount of 50 per cent of the total cost, it is extremely difficult for MoD or the Air Force to really work out where such large-scale offsets can be utilized to benefit the Indian Air Force.

Secondly, it is also difficult to cost an offset. We find that some firms offer us an offset for particular equipment and in another contract they are offering that they will upgrade that equipment as part of the offset of that contract. This makes costing a complex issue. Thirdly, whenever we tender out procurement, we get requests for deviation from request for proposals (RFP). We need to reach a mature stage before we decide which deviation can be accepted and which cannot be. Fourthly, if offsets are to be given in such large numbers, how do we actually ensure an audit? We need to be careful so that the proffered offsets are given and can be accounted for and the deals do not go awry. The last issue is that when we go about creating offsets on the technology side, we lack domestic vendors who can add value.

While looking at offsets, government cannot stop at merely making Indian industry a part of global supply chain of Defence Majors but must also seek system integration in India. The ToT agreements need to be restructured. The myth about the capability of the Indian industry to absorb the huge quantum of defence offsets need to be looked at from the consideration of capability and capacity and track record.

OVERSIGHT AND ACCOUNTABILITY

There are several areas where a financial perspective on offsets calls for care and caution. Some of these are enumerated here.

Value addition within the country is a cardinal principle of offsets and hence, where exactly the offset manufacturing is taking place is important.

In case of suppliers selected on competitive bidding, there may be an issue if the offset obligations are changed after contract. It will be difficult to verify whether change of offset has resulted into any financial benefit to the vendor and if so, the quantum of such benefit.

Even during selection of successful bidder, there could be operational issues in separate or concurrent evaluation of technical offers, offset offers and commercial offers.

Theoretically, the offset part of a quote should be treated at par with payment terms and be harmonized across all vendors before determining L1 because offset has a bearing on the quoted price. There is a very wide choice of items available for offsets and the list of options is proposed to be further expanded. Different vendors may respond with different offset portfolio. The L1 is determined by assuming that the offsets offered by all vendors are financially equivalent, which is not the case in reality. For complete harmonization of offsets offered by different vendors, perhaps the ideal course of action would be to specify the expected offsets in the RFP so that these are common to all bidders but then the flexibility available to vendors to choose from a big menu of options would be lost. Pre-specifying exactly what offset the buyer is looking forward could be considered, at least in selected high value cases.

The primary purpose of the offset policy should be increased indigenous production and access to critical technology. Indirect offsets should be secondary and need to be gradually rolled back as the industry's capacity to absorb direct offsets increases. In particular, investment-in-kind raises serious regulatory issues of non-competitive procurements as freebies with little accountability for valuation.

CONCLUSION

Value of defence acquisitions in the last seven years has exceeded the total amount of all acquisitions in all the previous years after Independence. The volume and costs of procurement and acquisitions on the capital side are whopping and would increase exponentially with many big ticket proposals in the pipeline. It has been estimated that in the 12th Five Year Plan alone, India expects more than US$10 billion (approx. ₹65,000 crore) to flow into the country through offsets. This means that on an average for every year of the plan, offsets worth ₹13,000 crore will have to be processed by Defence Offset Management Agency. Therefore, over the next five to ten years, as the policy matures, we must endeavour to make the most out of defence offsets. With a talented and educated workforce, India's IT, aerospace and defence industry has embarked on a journey to become a world-class indigenous manufacturer to meet India's current and long-term needs. We would like to aim at more and more system integration in India although the job creation potential of system integration is relatively lower. Increasing domestic value addition (contract manufacturing) through offsets can create more jobs but only lower end jobs. While access to less expensive workforce in India is a factor that can improve outsourced manufacturing, care has to be taken to maintain quality because such decisions on outsourcing of manufacturing are not driven purely by costs, but more importantly by the quality of workforce, work ethics and the systems of quality assurance/quality check. This applies both to elite engineers/managers and shop floor workers.

We expect the private sector to play an important role in product design, development manufacturing and integration capabilities available in the private sector to augment the capacities built in the public sector through PPP.

Imparting requisite knowledge, skills and attitude to the workforce is necessary to prepare them for newer jobs and skills. A lot of investment is required in on-the-job training, both induction training as well continuing education programmes, to sustain/ improve the quality of manpower.

The MoD regularly reviews and considers revisions to the offset policy as part of the DPP review process. The importance of evolving offset guidelines providing a stable regime cannot be overemphasized because related investments are huge and long term and, therefore, frequent policy changes would upset these decisions.

Purely from the finance perspective, it is felt that a strong audit mechanism is needed to enforce and verify compliance of offset obligations as false claims of compliance may go uncontested if there is collusion between foreign OEM and Indian offset partner. Audit/inspection of records of private sector offset partners would pose special challenge. There should be proper disclosures and arrangements on access to books and manufacturing facilities. The importance of this concern cannot be overemphasized because there have been cases in the past where export obligations attached with tax concessions were not met by the beneficiaries.

4

Objectives and the Potential of Offsets

Mrinal Suman

Abstract

Offsets are not freebies that a vendor offers to push his product. They are important for making decisions. Offsets are formal arrangements of trade under which a foreign supplier undertakes programmes to generate benefits for the economy of the buyer country. Most countries consider offsets as a leverage to obtain compensatory benefits either to fulfil an urgent economic need or to fill a technology gap. Offsets have to be in consonance with national economic objectives, as it is not the type of offset but its relevance that dictates the selection. Although India has been demanding offsets in defence deals for the last six years, it has still not streamlined its offset regime. Being one of the biggest buyers of defence equipment, India can draw immense benefits with a well-thought-out offset policy. Major aspects of India's defence offset policy have been appraised in the chapter to ascertain their relevance and applicability in the current environment.

INTRODUCTION

Offsets can generally be termed as formal arrangements of trade wherein a foreign supplier undertakes specified programmes with a view to compensate the buyer as regards his procurement expenditure and outflow of resources. In other words, the supplier undertakes programmes to generate benefits for the economy of the buyer country. It is a formal arrangement as it has inbuilt contractual obligations. The negotiated package consists of the primary contract and the compensatory offsets.

Offsets can also be called as trade arrangements with built-in reciprocity clauses to provide some sort of relief to the buyer to help him pay for the purchases. Perhaps, it would be best to define offsets as some sort of a leverage exploited by a buyer to obtain compensatory benefits by asking the seller to undertake well-designated activities to boost the economy of the buying country.

Although the World Trade Organization prohibits its signatories from imposing, seeking or considering offsets for government procurement transactions, offsets have become an integral part of world trade and are here to stay. Therefore, it will be prudent to consider offsets as a natural interplay of the market forces.

Earlier, arms exporters dictated terms by inflating prices and imposing other stringent conditions, which varied from demanding access to the importer's market to establishing military basis. Now, it is a buyers' market and they try to extract most profitable offsets to use them as engines of national economic growth, by redirection of large outflows involved in defence procurements back into their economy.

With limited buyers, the sellers are hard put to outbid their competitors. They have to make their offers virtually irresistible with promises of lucrative offsets. Generally it is for the buyer nation to decide as to what offsets to seek. It is a very crucial decision and demands careful consideration. It is not the type of offset but its relevance that should guide the selection. Therefore, offsets should be in consonance with the national economic objectives. Value of an offset is always expressed in terms of percentage of the value of the main contract.

MANAGEMENT OF OFFSET PROCESS

The success of any offset programme primarily depends on proper selection, detailed planning, close supervision and regular monitoring. Therefore, the whole process of offsets has to be managed in a well-thought-out and coordinated manner.

The complete offset process can be broadly considered in five stages:

1. **Policy stage.** This is the initial and perhaps the most important stage. Offsets should form a part of the overall national endeavour with well-specified aims. A national offset policy needs to be formulated with the objectives that are sought to be achieved through offsets, clearly spelt out. The policy statement should also lay down offset thresholds and indicate the areas in which offsets are preferred. This helps the sellers in preparing their offset packages.

2. **Planning stage.** This stage starts with the receipt of bids from the sellers, which are evaluated along with their offset packages. Discussions are carried out for seeking clarifications. Once the successful bidder is identified, a detailed dialogue is initiated with him to draw out a mutually acceptable offset plan, which is flexible, realistic, realizable and practical. The plan specifies the sectors in which offset programmes are to be implemented with inter-se priorities and assigns multiplier values to them.

3. **Negotiation stage.** Specific projects in each sector are identified. Various options to have an optimally balanced mix are negotiated. Expert groups are constituted for different projects and their reports included in the contract document. All aspects are covered in clear and unambiguous terms such that interpretation would not lead to any confusion. It is imperative that the seller is bound by various clauses to prevent his reneging on his promises. Therefore, levels of technology, value addition, penalty clauses, measurement methodology and time frame for implementation are spelt out clearly in the contract document. While being specific, there generally is enough inbuilt flexibility to cater for changing conditions.

4. **Implementation and monitoring stage.** Implementation of offsets is invariably a long-drawn process. Its success

depends on sincerity in execution and strict adherence to the letter and spirit of the agreement. A properly constituted monitoring mechanism is put in place to carry out periodic reviews of the process and apply corrections, where necessary.

5. **Feedback and review stage.** It is really a stage of doing stock-taking. A thorough review of the whole offset programme is carried out to ascertain the degree to which the stated objectives have been achieved. Weaknesses and infirmities are identified and corrective measures recommended for future. This stage provides vital inputs for the periodic review of the overall offset policy as well.

Offsets do not come for free and entail a cost penalty. It is generally seen that offsets up to 50 per cent inflate the cost of the main contract by close to 10 per cent. Similarly, 100 and 200 per cent offsets may result in cost escalation by 15 and 20 per cent, respectively (see Figure 4.1).

Figure 4.1 shows the following:

(a) Cost penalty is not directly proportional to offset percentage. Cost penalty varies with the type of programmes undertaken.

(b) Initial cost of establishing infrastructure and initiating offset programmes is considerable. Once the process becomes functional and gets streamlined, rate of increase in cost penalty drops.

(c) Cost penalty tends to plateau after offset percentage reaches 200 per cent. It implies that beyond this stage, both the offset provider and the receiver remain engaged in programmes that make commercial sense to both.

To obtain maximum benefits from offsets, all countries spell out objectives of their offset policies at the outset and thereafter formulate their policies to achieve them. Therefore, offsets are justified only when the advantages accruing from them outweigh

Figure 4.1 Relationship between offset percentage and cost penalty

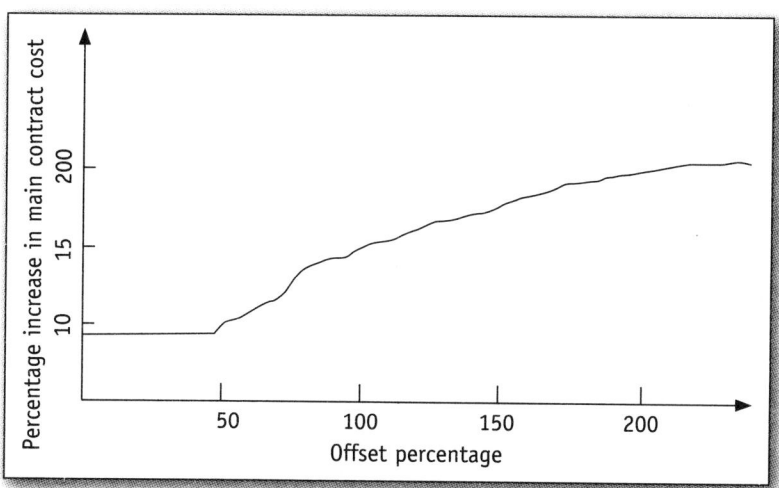

the related overheads. This is the overriding principle and should never be forgotten.

POLICY WITHOUT OBJECTIVES

Over 120 countries are demanding offsets in one form or the other. Offset policies of all nations start with the aim clearly defining as to what that country wants to achieve through this route—it could be upgradation of technological industrial base, import of dual use technology, development of small and medium enterprises, generation of employment opportunities, improvement of competitiveness of local defence industry and seeking new markets for defence products. So every country spells out what it actually wants to achieve through offsets.

Every country has a national offset mission or policy. All other policies, say policy for civil aviation and policy for defence, flow from that. India has no national offset policy. The Defence Offset Policy is a stand-alone and exclusive initiative of the Ministry of Defence (MoD).

During the period of 2004–2005, annual defence exports were a paltry US$50 million, whereas our imports were in billions of dollars. It was decided to make use of the mechanism of offsets to persuade foreign vendors to buy goods and services from the Indian defence industry, meaning the public sector. Initially, announced in 2005, the policy was totally loaded in favour of the public sector to the extent that even the task of monitoring implementation of offsets were assigned to it. It was more of a counter-trade arrangement designed primarily to promote exports from the public sector.

Under pressure from the private sector, MoD expanded the scope of offsets in 2006 to include any private defence industry manufacturing these products or components under an industrial licence granted for such manufacture. Subsequently in 2008, the mandatory requirement of an industrial licence for offsets for private companies was also removed and offset banking introduced. In January 2011, the scope of offset activities was extended to include civil aerospace and internal security sectors. Training services and equipment were also included but not training infrastructure.[1]

Thus, Indian policy is a product of evolution in bits and pieces. It does not owe its origin to a well-thought-through strategy. The policy has no spelt-out aims. Today, nobody in the country (or outside) really knows as to what India wants to achieve through offsets. The procurement procedure proclaims self-reliance to be one of the objectives. However, the offset policy does not contribute towards the said objective.

India allows three routes for the fulfilment of offset obligations.[2] These are

(a) Direct purchase of or execution of export orders for, eligible products and components manufactured by, or services provided by, Indian industries, i.e. Defence Public Sector

---

[1] Editors' note: Since then, defence offset guidelines have been released in June 2012 and the DPP 2013 announced in April 2013. A DIPP proposal is also in circulation as of 2014 for major changes possibly in 2015.

[2] Editors' note: In defence offset guidelines released in June 2012, additional options for discharge of offset obligations were made available.

Undertakings, the Ordnance Factory Board and private Indian industry.

(b) Foreign direct investment (FDI) in Indian industries for industrial infrastructure for services, co-development, joint ventures and co-production of eligible products and components.

(c) FDI in Indian organizations engaged in research and development (R&D) as certified by Defence Offset Facilitation Agency.

Products eligible for discharge of offsets relate to defence, internal security and civil aerospace. 'Services' mean maintenance, overhaul, upgradation, life extension, engineering, design, testing of eligible products and related software or quality assurance services with reference to the indicated eligible products and training. Training may include training services and training equipment but exclude civil infrastructure.

An examination of the viability of the above three routes is very revealing. At the face of it, FDI route appears highly viable but when considered in the light of upper cap of 26 per cent, the dissuasiveness of the policy becomes apparent. A foreign investor is expected to invest his resources in a venture where he has no significant control, strict capacity/product constraints, no purchase guarantee and no open access to other markets including exports. The FDI policy was announced in May 2001 and has elicited no response. Therefore, options (b) and (c) above mean little.

Export of defence goods and services is thus the only possible method to fulfil offset obligations under the Indian policy. It amounts to compensation trading.

DIRECT PURCHASE OF PRODUCT AND SERVICES

Export of goods and services is considered to be the least beneficial form of offsets. It provides temporary and illusory gains, as has been the experience the world over. New markets dry up soon after the offsets are fulfilled, leaving the infrastructure created idle and bereft of orders. Gains of counter-purchase are always transitory. Therefore, such trading does nothing to either strengthen the defence industrial base or promote technology upgradation. In other words, a potent tool like offsets is wasted on one-time exports as they do not boost the economy of the buying nation.

Another issue of concern relates to Indian industry's capability to absorb billions of dollars worth of offsets by exporting equivalent goods. It is felt that a nation whose defence exports have never exceeded US$50 million can never take such an enormous leap. Inclusion of civil aerospace and internal security sectors means little as these are analogous fields and the manufacturers remain the same. For example, both for civil aviation and combat aircraft, only HAL possesses the necessary infrastructure. The same is applicable to internal security sector—the manufacturers remain the same; only the scope has been enlarged.

FDI IN INDIAN DEFENCE INDUSTRY AND R&D

While opening the defence industry to the private sector in May 2001, the government allowed 26 per cent FDI. It was hoped that foreign investors would rush in with their bags of money. However, all hopes have been belied and the policy has been acknowledged as a total failure. Most prospective foreign investors view the policy to be highly dissuasive in intent and content. There has been a total lack of enthusiasm on the part of foreign investors.

A cap of 26 per cent is considered to be highly unreasonable as foreign investors get no significant control over the enterprise. Additionally, the policy mandates that the Chief Executive has to be a resident Indian and management control must remain in Indian hands with majority representation in the board. Many policy provisions are perceived to be highly restrictive—a licensee can produce only the licensed products and in the sanctioned quantity; he can neither diversify nor enhance production; the government can give no purchase guarantee; and the licensee will have no open access to other markets, including exports.

As India needs FDI in defence primarily for technology infusion, it should adopt a flexible and proposal-specific approach. All joint venture proposals should be assessed and categorized as follows on the basis of nature, level, depth and exclusivity of technology involved for the fixation of FDI cap. Defence is too vast a sector to have a single cap. The following technology-centric criteria should be applied:

Tier 1: Low-tech proposals—up to 26 per cent
Tier 2: High-tech proposals—up to 49 per cent
Tier 3: Proposals with latest technologies—up to 74 per cent
Tier 4: Proposals with cutting-edge frontier technologies—up to 100 per cent

USING OFFSETS FOR INFUSION OF TECHNOLOGY[3]

Self-reliance is a stated aim of Defence Procurement Procedure (DPP) 2008. Yet, India does not accept technology under offsets. It is a dichotomy. As seen above, offsets cost a country considerable extra expenditure. It is universally accepted that offsets make sound business sense only if the trade-off results in extraordinary economic or technological gains. Technologies that industrially advanced countries are reluctant to sell can only be obtained through the leverage of offsets.

Transfer of technology (ToT) is the most common type of direct offset in defence transactions. Up to 30 per cent of all offsets provided relate to technology transfer of varying degrees. Most of the buyers want the technology to manufacture the complete system, if possible, or at least its sub-assemblies.

Genuine ToT implies transfer of know-how to enable the recipient to produce equipment from component and raw material level. Selection of technology should be need based and not availability

---

[3] Editors' note: Transfer of technology under defence offsets is now an option in DPP. The paragraph below should be read in this light. However, the points mentioned by the author remain pertinent.

driven. It should fill a critical gap and provide a platform for the development of more advanced technologies. It should have multiple applications for economies of scale.

A successful ToT depends on four essentials—careful identification of technology for import, selection of reliable OEM, objective nomination of recipient and close oversight of technology absorption. ToT must fill critical gaps in indigenous knowledge and help accelerate the process of achieving self-reliance. Therefore, its selection should be need based and not availability based. Further, technology should be of the latest genre with multiple applications.

Selection of suitable OEM is the key to successful implementation of a ToT contract. In addition to his capability to transfer the necessary know-how, OEM should have a track record of honest and diligent adherence to contractual obligations. Unreliable and unscrupulous OEM must be conscientiously avoided. It should also be made certain that there are no strings attached to the said technology as regards its unrestricted use by the recipient.

A technology is a combination of a large number of sub-technologies. Once a technology package is broken down into sub-technologies and duly categorized, a market scan should be carried out to identify the entity (private or public sector) most suited to receive and absorb incremental sub-technologies to be able to produce the equipment. Such a course of action will not only cost least but will also expedite indigenous production. Unless fully absorbed and mastered, ToT can prove highly wasteful. Therefore, it is absolutely imperative that the complete process of transfer and absorption is closely monitored by an independent agency.

LEVERAGING PURCHASING POWER FOR OPTIMUM GAINS

Value of an offset depends primarily on its appropriate selection. Ill-conceived and ill-planned offset programmes invariably prove to be highly wasteful in national resources and uneconomical for their value. Therefore, programmes have to be selected on the basis of their viability, estimated offset credit value, ease of monitoring and demonstrability of accruing benefits. Offsets should not be viewed in isolation as one-time agreements, but as an important and integral element of long-term national policy. To derive full benefit from offsets, it is absolutely necessary to understand the dynamics of offsets. Being one of the biggest buyers of defence equipment, India can draw immense benefits with a well-thought-out offset policy. It is time India puts its act together for optimum exploitation of offsets in all defence deals.

Offsets are commonly assigned 'multiplier value'. It is a factor applied to the actual value of an offset transaction to calculate the credit value earned. It is a methodology of assigning weightage to different offset programmes. Buyers use multipliers to provide sellers with incentives to offer offsets in targeted area of their choice.

Let us say there is a third-generation technology for night vision devices and the cost of that technology is US$10 million. India can offer a multiplier of two. It would imply that the seller would offset credits worth US$20 million. In case fourth-generation night vision technology carrying a multiplier value of 4 is offered, the seller will get offset credits worth four times the cost of technology. It is a highly effective way of channelizing offsets into the areas of buyer's choice

PRICING OF TECHNOLOGY UNDER OFFSETS

The common reason given by the Government functionaries for non-acceptance of technology against offsets is that they are unable to price technology. It is a totally specious and hollow excuse. India purchases technology as a part of 'Buy and Make' deals—either contracted as a part of the main deal itself or negotiated separately at a later stage. Apparently, no ToT contract can be entered into without pricing the technology involved.

There is no standard price of technology in the world market. It is purely need based and determined by the degree of desperation of the technology seeker. Defence Research & Development Organisation (DRDO) can be asked to identify technologies that it needs.[4] Normally, DRDO should seek technologies in which it has made considerable headway but not mastered—commonly referred to as lacking 'last mile connectivity'. Thereafter, DRDO can work out the anticipated expenditure on indigenous effort to master the said technology as also urgency of its application. These two factors help decide fair, reasonable and acceptable price of technology.

Many knowledgeable observers are of the opinion that MoD has been pressurized by the public sector to disallow ToT against offsets as the public sector wants to perpetuate its monopoly on receipted technology through the current system of 'Buy and Make' deals. The public sector fears that acceptance of ToT against offsets would make the private sector preferred partner of foreign vendors. Therefore, opposition of the public sector is purely to safeguard its exclusive turf. It is time the MoD rises above extraneous consider-ations to ensure that the benefits accruing from offsets do not get outweighed by the cost penalty. ToT should be made the preferred form of offsets. Through a well-evolved system of application of multipliers, irresistible incentives should be provided to foreign vendors to offer technologies that India needs the most.[5]

CONCLUSION

Presently, offsets are neither contributing to the upgrading of indig-enous technological base nor reducing dependence on imports. Indian policy makers have failed to appreciate the full potential of offsets and have trivialized a powerful instrument of industrial growth for transitory gains for a few public sector entities. It will not be incorrect to state that the Indian offset policy is designed to obtain export orders for the public sector at the cost of the defence budget.

Finally, offsets should not be viewed in isolation as one-time agreements, but as an important and integral element of long-term

---

[4] Editors' note: This aspect is included in DOG 2012.
[5] Editors' note: ToT and multipliers have been incorporated in DOG 2012.

national policy. To derive full benefit from offsets, it is absolutely necessary to understand the dynamics of offsets. Being one of the biggest buyers of defence equipment, India can draw immense benefits with a well-thought-out offset policy. It is time India puts its act together for optimum exploitation of offsets in all defence deals.

5

Defence Offsets and Capability Build-up

S.N. Misra

Abstract

The author discusses four major issues related to offsets in this chapter. These are transfer of technology, policy disconnects, direct and indirect offsets and capability build-up. He has made extensive use of data to drive home his points. He opines that critical technology in the global market is not priced economically but strategically. Offsets have a major role to play in obtaining this technology. Further, there is a policy disconnect within various ministries in the government that needs to be bridged. He is a strong proponent of inclusion of commercial shipbuilding in the ambit of indirect offsets and believes that the key to a successful implementation of offsets importing know-why and not know-how.

INTRODUCTION

The amount of data that is available in public domain on offsets is either incomplete or sketchy. The international position is not far better except for the US where the Department of Commerce brings out comprehensive report on all offset transactions from an arms suppliers' viewpoint. This makes it difficult to do any meaningful study on impact of offset on building military industry capability, bringing technology, etc. There are four major issues related to offsets that are discussed in this chapter: technology transfer, policy disconnects, direct and indirect offsets and capability build-up.

TRANSFER OF TECHNOLOGY

When we talk about technology transfer, the experience over the globe has been that mostly obsolete technology has been transferred; only the degree of obsolescence varies. The issues most germane to technology transfer pertain to the depth and range of the transfer. For instance, in the aircraft industry, which is a major player in offset realization, technology being transferred will not be current due to export control regulations, commercial and intellectual property interests in the technology developed. The domain knowledge in manufacturing invariably lies with the tier suppliers and it has been HAL's experience that global players provide only telescopic information on queries and clarifications that are sought on the technologies while adapting them to our environment. This results in project delays and time overruns.

Documentation and configuration control is another area where there will be several gaps during transfer of technology (ToT). Documents supplied are generally from the archives of the licensor and need lot of upgradation during the production phase. Process sheets are not transferred in their entirety as the original equipment manufacturer (OEM) depends on his vendors for a large proportion of the work who hold these documents. Design documents with details of stress analysis, dynamic and thermal loads, ballistic profiles, etc., are not transferred. This would mean that dependence on OEM would continue for concessions, modifications, tropicalizing and upgrading unless the basic design knowledge is available with the licensee.

While manufacture of components and sub-systems could be undertaken through a process of ToT with relatively less number of issues and problems, system integration in an aircraft is a different ball game. To give a typical example, aircraft flight control system integration involves a myriad of complex sub-systems (such as Hydraulic, Pneumatic, Electrical, Electronics and Fuel) that demand sound domain knowledge in complicated fields of aero-dynamics, flight mechanics and aero elasticity. Complex interface issues arise during the integration process for which solutions are not clearly defined in the ToT document.

The other issue that merits serious attention is the quality of technology being offered and evaluation of technology. Scholars say there are costing methods like the income method and the investment method. In actual terms, however, there is no comprehensive evaluation of technology transfer anywhere in the world and tools for valuing technology are inherently inept. But what one can certainly do is to flag the critical technology that one requires and identify the potential countries from where they can be sourced and commit them to pass on the capability to manufacture critical sub-systems with maximum range and depth.

An effective way to obtain these critical technologies is by leveraging our big ticket acquisitions. It is unlikely that it will come unless we have strategic partnership as we had with the erstwhile USSR and presently Russia. If the US offers us AESA technology (Advanced Electronically Scanned Array Radar), which is required for all frontline fighters such as Sukhoi 30, Mirage and FGFA, it would be only given with a tacit understanding that India is going to buy the aircraft from the Americans. Competitive tendering is unlikely to give India any strategic technology. Strategic alliances would only give us the kind of technology that we require, as South Korea got from the US.

It would be useful to identify gaps in some of the critical technology that we need badly to shore up our indigenous military capability. Table 5.1 delineates some of the gaps, sub-systems wise.

DIRECT AND INDIRECT OFFSETS AND POLICY DISCONNECTS

An important aspect in offset policy is the need of a more effective mix of indirect and direct offsets. Most of the countries do have such a mix that benefits both civil and defence sector bringing in

Table 5.1 Critical technology gaps

S. no.	Systems	Gaps
1.	Gas turbine engine	Single crystal and special coating in turbine blades, FADEC
2.	Missile	Uncooled FPA Seekers ·
3.	Aeronautics	Smart aerostructures and Stealth technology
4.	Material	Nano-material and carbon fibres
5.	Naval systems	Super cavitating technology
6.	Sensors	AESA, Radar, RLG and INGPS
7.	Communication	Software-defined radio
8.	Avionics	Gen III, II tubes
9.	Surveillance	UAVs and Satellites

complementarities and synergy for macro-economic development. Details of such a direct/indirect mix for 20 countries are given in Table 5.2.

It would thus be seen that 60 per cent of the countries opt for both direct and indirect offsets, while the rest target defence specific or civil sector benefits. In the US, during 1993–2009, the total contract value of offset was of the order of $76 billion with 46 countries. Of these, 37 per cent was direct and 63 per cent indirect. Eriksson (2007), in his study of European Defence Industries during 2000–2005, brings out that offset worth $4.2 billion was concluded of which direct offset was 40 per cent, while indirect defence offsets and civil indirect offset were 35 per cent and 25 per cent, respectively.

As per the 12th Plan (2012–2017), India's total fund requirement for infrastructure development will be around $1,025 billion.[1] It has been proposed that such massive funding requirement should be

---

[1] Eriksson IA, Study on the Effects of Offsets on the Development of European Defence Industry & Market, 2007. Available at: http://www.eda. europa.eu/docs/documents/EDA_06-DIM-022_Study_on_the_effects_of_offsets_on_the_Development_of_a_European_Defence_Industry_and_Market_1.

Table 5.2   Offset policies of select countries

S. no.	Country	Minimum value of defence contract	Minimum offset required	Offset sector
1.	Australia	US$3.75 million	No specific min. or max.	Defence
2.	Canada	–	100%	Defence & Civilian
3.	Finland	€10 million	100%	Defence
4.	Greece	€10 million	120%	Defence
5.	Israel	US$0.5 million	35%	Defence & Civilian
6.	Italy	US$6.6 million	Not less than 70%	Defence
7.	Netherlands	€5 million	100%	Defence & Civilian
8.	Norway	US$6.7 million	100%	Defence & Civilian
9.	Poland	€5 million	100% (min. 50% defence)	Defence & Civilian
10.	South Korea	US$10 million	30%	Defence
11.	Spain	NA	100%, but may vary	Defence & Civilian
12.	Switzerland	US$17 million (may vary)	100%	Defence & Civilian
13.	Taiwan	US$10 million	70%	Defence
14.	Turkey	US$10 million	50%	Defence & Civilian
15.	UK	US$17.2 million; £50 million for French & German companies	100%	Defence
16.	Austria	$1 million	100%	Defence & Civilian
17.	Brazil	$5 million	10%	Defence & Civilian
18.	South Africa	$2 million	50%	Defence & Civilian
19.	UAE	$10 million	60%	Civilian
20.	Germany	$5 million	Negotiable	Defence & Civilian

SOURCE: Adapted from U.S. Department of Commerce, Bureau of Industry and Security, 'Offsets in Defence Trade: Twelfth Report to Congress', December 2007.

provided through 50:50 public partnerships as against 70:30 that was the funding pattern during the 11th Plan. This entails massive investment in different core industry sectors. The growth trend in core industry sectors is given in Table 5.3.

While some of the core industry sectors like telecom and roads have witnessed handsome growth, the social sector in India really lags behind with India ranking 119 out of 169 countries with a score of 0.519 in the Human Development Index. The trend in social sector expenditure as a percentage to central government expenditure and GDP is indicated in Table 5.4.

It would be seen from the above tables that allocation to social sectors remains abysmally low (around 9 per cent) compared to developed countries that spend close to 20 per cent of their GDP on education and health. In India, about 22 per cent of the population lives below subsistence level (i.e. the poverty line) with very little

Table 5.3 Core industry growth rates

Sector	2007–2008	2008–2009	2009–2010	FDI inflow 2009–2010
Power	6.3	2.5	6.8	$1,237 M
Coal	6.0	8.2	8.0	–
Steel	6.8	13.2	3.2	–
Railways (freight)	9.0	4.7	6.6	$25 M
Cargo at ports	12.0	2.2	3.7	$65 M
Tele-communication	38.3	80.9	47.3	$2,223 M
Fertilizer	8.6	2.6	13.2	–
Cement	7.8	7.6	10.1	–
Petroleum	0.4	–1.8	0.5	$218 M
Civil aviation	11.9	3.8	5.7	$16 M
Roads	164.6	30.9	21.4	–

SOURCE: Ministry of Finance, Government of India, Economic Survey 2010–2011(P/289).

growth in employment in the organized sector (−0.65 per cent in public sector and 1.75 per cent in private sector) during 2009–2010.

The massive investment needs in infrastructure need to be complemented with investment in human resource development and inclusive growth. The offset policy must include social sector as part of indirect offset policy as many countries have done.

The offset policy in 2009 included civil aerospace sector in the ambit of offset. However, the commercial shipbuilding has been left out of the purview. Countries such as South Korea and China have built significant capability in shipbuilding as Table 5.5 illustrates.

Table 5.4   Central government expenditure in social sectors (%)

Social service	2007–2008	2008–2009	2009–2010
Education	4.02	4.04	3.96
Health and family welfare	2.05	1.91	1.90
Water supply	2.02	2.30	2.20
Social welfare nutrition	0.80	0.70	0.80
Total	8.89	8.95	8.86

SOURCE: Ministry of Finance, Government of India, Economic Survey 2010–2011.

Table 5.5   Share in world's order for shipbuilding

Country	2004	2009	2010
Japan	34	17.3	16.3
South Korea	37.2	34.7	34.3
China	14.0	37.0	39.5
Europe	5.2	2.2	1.6
Others	0.2	1.1	0.9

SOURCE: HIS (Former Lloyd's Register) 'World Shipbuilding Statistics' year end (Shipbuilding Statistics: Shipbuilding Association of Japan—March 2011).

Table 5.6 Multiplier by civil aviation sector

Type of transaction	Multiplier
Training in India	5
Training abroad	2.5
Technical assistance	2.5
Investment in joint venture	1.5

One of the policy facilitators for their global footprint is their offset policy and targeting technology and training. Commercial shipbuilding and warship building have to be looked at in unison as beneficiary of offsets.

However, when we wish to open up defence offsets to indirect sectors, there is a visible disconnect between various sectors. Recently, the aviation sector was opened up to the ambit of defence offsets. One of the serious problems is that there is hardly any interaction between the Defence Ministry and the civil aviation sector. It would be a surprise to know that a multiplier policy exists in the civil aviation sector; they do give multipliers varying from 5 to 1 depending on the level of the benefit that they are obtaining, whereas in defence multiplier is non-existent (Table 5.6).[2]

Even if a policy of providing offsets to the aviation sector is implemented, it will not work until the civil aviation ministry and the MoD work in tandem. There has to be better synergy between the defence ministry and the civil sector.

CAPABILITY BUILD-UP

In India, the research and development (R&D) investment in the defence sector predominantly comes from Defence Research & Development Organisation (DRDO). The inadequate R&D

---

[2] Editors' note: Defence multipliers have been introduced in the Offset Policy of 2012.

investment in India would be evident from the global comparison given in Table 5.7.

Further, Table 5.8 would show global leadership in terms of various parameters of R&D.

It would thus be seen that though India ranks first in terms of number of qualified engineers it lags seriously in terms of skill, qualitative scientific institutions and investment in R&D.

Table 5.7 Defence R&D expenditure global trend ($B)

Countries	R&D exp.	Per cent to total mil exp.
USA	90	14%
Russia	7	11.5%
France	6.1	11%
UK	4.7	9%
India	2.0	6%

SOURCE: Keith Hartley, 'Defence R&D: Data Issues', *Defence and Peace Economics*, Vol. 17, No. 3, June 2006, pp. 169–175.

Table 5.8 R&D leadership against various parameters

S. no.	Parameter	1st	2nd	3rd	4th
1.	R&D exp. % GDP	Israel	Sweden	Finland	France
2.	Skilled labour	Denmark	Iceland	Israel	Austria
3.	Qualified engineers	India	Finland	Israel	Japan
4.	Technology readiness	Israel	USA	Finland	Sweden
5.	Quality of scientific institutions	USA	Sweden	Israel	Finland
6.	Utility patents	USA	Japan	Taiwan	Israel

SOURCE: *World Competitiveness Year Book 2005*, Ministry of Defence.

Further, a survey of the R&D investment in India reveals that the private sector invests very little in R&D. For instance, 0.5 per cent is invested by Larsen and Toubro on R&D. Tata has a budget of 0.7 per cent as the R&D investment. With such scant priority to R&D, the capacity to absorb critical technology and manufacture major sub-systems would be grossly inadequate. The lesson learnt is that just by paying for technology transfer and entering into licence production you are not going to have technology unless you have the R&D readiness and wherewithal in this country. One can build capacity; a company can build a huge dry docking dock but yet not be able to build a frigate. For building the frigate, capability is required. For capability build-up there is a need to have design capability, public–private partnership both in terms of production, joint venture (JV) with OEM, joint technology development with major design houses and significantly higher R&D investment by Indian industry. The key is know-why and not know-how.

THE ROAD AHEAD

India embarked upon the liberalization path in 1991 by dismantling the licence, quota permit raj. In the defence sector, the wave of liberalization came a decade later bringing in private sector as partners and allowing 26 per cent foreign direct investment (FDI). However, the liberalization policy in defence and offset policy is so far rather half-hearted and myopic and implementation arrangements under Defence Offset Facilitation Agency (now DOMW) extremely inept. The government needs to develop a comprehensive industrialization strategy where defence industry is part of the larger policy mosaic to upscale our manufacturing capability as China has done. Use of offsets can be a critical facilitator of this strategy.

Of the four BRIC countries—Brazil, Russia, India and China— that are expected to be global leaders in 2020, Brazil, China and Russia can be seen to have taken significant strides in aerospace, shipbuilding, manufacturing and exports with offsets playing a significant role. India is yet to realize its full potential in these areas.

In this regard, FDI should be increased to at least 50 per cent as this is likely to bring in significant, key manufacturing and design technology capabilities. Simultaneously, substantial R&D investment and joint technology projects will be needed to spur export potential and provide spin off to the civil sector through dual end technology. Front-end defence technology in electronics, avionics and metallurgy and propulsion has always been a precursor to overall growth in S&T and national capability build-up. To trigger this process higher social sector investment, skill upgrading and training are inescapable. A mix of direct and indirect offset will facilitate this process. India must follow growth models of countries like Japan, Turkey, Brazil, Israel, China and Finland who have seen significant success. A liberal offset policy with strong political will have the heady potential to make India a hub in defence global supply chain along with helping us upscale our self-reliance quotient.

CONCLUSION

We must target critical technology, we must try to value-add rather than integrate and we must invest more in R&D. The public sector bashing has to stop. The way forward for the country is public–private partnership, JVs with reputed OEMs and technology co-development with major design houses. Government has to ensure a level playing field between the private and the public sector and do away with the preferential treatment to Defence Public Sector Undertakings. The acrimony between the public and private sectors, the DRDO and the users is not going to help the country. Instead of this mutual bashing and turf war between the services, public sector, private sector and DRDO, it is time that there is political will to mentor and ensure proper synergy between these entities.

6

Chinese Takeaways for Building a High-tech Defence Innovation Base in India

Smita Purushottam

Abstract

There is a strong, established correlation between a country's investment in research and development and its international competitiveness. A comfortable balance of payments position, a strong currency and high living standards are attributes of a high-tech economy.[1] A big economy usually reaches this stage after undergoing an industrial revolution that has produced a strong manufacturing sector, which is a prerequisite for supporting a viable defence industry. While the US is the ultimate example of a successful innovation model, China's experience may be more relevant for India, as both countries face similar challenges as developing countries trying to liberate themselves from earlier autarkic, State-led and low-tech paradigms. This chapter attempts to outline the main arguments in this regard and suggests some solutions for establishing a high-tech manufacturing sector that can support a viable defence industry in India.

---

[1] Michael Porter, *The Competitive Advantage of Nations* (New York: Free Press, 2008).

INTRODUCTION

The shocking revelation in a 2011 Indian Ministry of Commerce strategy report that India's merchandise trade deficit would reach an unsustainable 11.5 per cent of GDP by 2014[2]—highlighted both India's depleting competitive advantage across a range of sectors and the need to address the root causes on an urgent basis.

Today, 65 years after Independence—India is still importing between 70 per cent and 100 per cent of its machine tools, telecommunications, electronics and aviation equipment, on top of massive commodity imports. The import content of some of India's 'indigenous' defence production items ranges up to 95 per cent and India imports nearly all its advanced weaponry from abroad. India is also importing high-end consumer goods including processed food, and consuming higher educational services worth billions of dollars instead of converting its own education sector into a world-class foreign exchange generator. The bigger tragedy for India's masses is that India is not only importing high-tech equipment, it is ceding ground in export markets to countries like Vietnam and Bangladesh in traditional, labour-intensive manufacturing.[3] India's growing requirements of manufactures and services thus benefit manufacturers and providers abroad while avenues for dignified employment in the economy shrink due to years of 'jobless' growth. India's services sectors are generating jobs for its young people as modern-day 'coolies', waiting on the rich at hotels, clubs, restaurants, beauty shops and malls, with little prospect for personal advancement. Meanwhile, India's elite is largely untouched as yet from these unfolding tragedies and does not even understand the danger it faces from the social discontent this lop-sided growth generates.

---

[2] 'Strategy for Doubling Exports in Next Three Years (2011–2012 to 2013–2014)', Government of India, Ministry of Commerce & Industry, Department of Commerce.

[3] Jonathan Anderson, Emerging Advisors Group, 'And the Three Reasons India Will Fade Away'. Available at: http://emadvisorsgroup.com/about-us.html, accessed on 24 November 2014.

In 2012, the sharp depreciation of the Indian rupee, which overnight shaved 20 per cent off our GDP—an outcome to be expected given long persisting current account deficits—only reinforced the need for urgent, well-thought-out reforms.

It is clear that only a strong, diversified manufacturing sector *and* autonomous technological capabilities generated by effective research and development (R&D) investments, which satisfy at least some of our requirements of both high-tech and labour-intensive products, can put the economy on a truly sustainable path and create the minimum conditions for building a strong defence industry, essential for national security. Further neglect would be another 'blunder of Himalayan proportions' according to a patriotic, senior government official. There is an urgent need for a multi-purpose high-tech manufacturing strategy that could guide decision-makers to reach these inter-related goals.[4]

DEFENCE SECTOR REFORMS PRESENT AN OPPORTUNITY

A beginning has been made by setting up the National Manufacturing Competitiveness Council. But ongoing reforms in defence procurement[5] and defence offsets also offer an extraordinary opportunity for rejuvenating industrial production and catalyzing economic growth. Though growth under this route will be led by the defence (and dual use) sector, a holistic, inclusive and well-thought-out strategy can energize and locate competitive, high-tech production across a range of civil and military sectors in India. *There are pitfalls in an exclusive reliance on growth led by*

---

[4] While I have regrettably not focused on the agricultural sector, a strong manufacturing sector has to be complemented by a healthy rural sector for a truly balanced Indian economy; but this can be built on the basis of nation-wide cooperative movements, micro-finance and infrastructural investments to address the abysmal performance on a range of social indicators.

[5] The Indian Ministry of Defence's Defence Production and Defence Procurement Policy reforms in 2011.

the defence sector, as pointed out in my earlier article,[6] the major ones being the danger of entrenchment of a defence oligarchy and crony capitalism, lop-sided economic development and increased dependence on foreign technology providers unless effective technology transfer provisions are built into a sound defence industrialization strategy.

Indeed, looking at the defence industrialization strategies of a number of countries, one can discern a clear trend of leveraging defence offsets to kick-start growth in the high-tech civilian and defence sectors.[7]

But why China as a model? I started looking at China when I found in 2008 that even Europeans and Americans were fretting about China catching up with them in science and technology (S&T). By 2009, China had become the second largest spender in the world on R&D after the US, out of an 'estimated $1.28 trillion in global R&D'.[8] China's R&D funding had grown at an annual rate of 20 per cent over the last decade.[9] In his 2011 State of the Union address, President Obama called the challenge of remaining at the cutting edge of global innovation—*America's new Sputnik moment*. The US–China Economic and Security Review Commission was charged with tracking China's progress in S&T and its implications for America's competiveness and national security. A recent Jane's report mentions an exponential increase in the number of patents given to China's defence industry. Clearly, it was time for soporific India to start paying attention to this phenomenon.

---

[6] Smita Purushottam, 'Will Defence Industrialisation Help the Technological Upgrade of the Indian Economy?' Available at http://www.globalpolicyjournal. com/blog/24/03/2011/will-defence-industrialization-help-technological-upgrade-indian-economy.

[7] 'The changing role of offset in the global defence market', Jane's *Industry Quarterly*, 3 November 2009.

[8] 'Science and Engineering Indicators 2012', National Science Board, United States of America.

[9] Ibid.

WHY STUDY CHINA'S TECHNOLOGY MODEL 2-3-4-5?

China is not the only model we need to study before deciding on the shape of our own national innovation ecosystem, which should be unique. Additionally, both the Chinese economic and S&T models have their fair share of problems and deal breakers, stemming from the hybrid nature of China's political economy. But I have warned earlier that we should not blindly imitate policies, even if highly successful, of other countries.[10]

Thus, the US is the most successful example of innovation and we can learn a lot from the institutions, policies and frameworks for encouraging science and innovation there. For example, the personal direction by President Obama of science policy and implementation through meetings of the President's Council of Advisors on S&T is something we should emulate (and have recommended in VISTAS—see below).

But China and India were, till the 1970s, following similar economic policies, causing the same distortions in development outcomes, a point I have made before. They are therefore now faced with similar challenges in trying to overcome these legacies, and so in some ways China's experiences are more relevant to us from the point of view of development transition models. Justin Yifu Lin, Chief Economist, World Bank and then Director of Beijing's University China Economic Research Centre had, in a beautiful monograph handed personally to me, lucidly outlined the parallels between the economic trajectories followed in the pre-reform eras of the two countries. Central planning, high exchange rates, discouraging exports, resources diverted into heavy engineering sectors, suppression of the private sector and the comparative advantages of their respective economies, and excessive dirigisme led to tragically deficient growth rates and sub-optimal social and economic outcomes. Both countries struggled with the legacy of a statist model and a low-tech base; both attempted the transition

---

[10] Smita Purushottam, 'Copycats don't catch mice', *Financial Express*, 18 November 2002.

from central economic planning to a market (and for China a high-tech) economy and hence can learn from each other.

A third reason for studying China was because it got its basics right when it decided to strengthen its infrastructure and manufacturing sectors and subsequently prioritize high-tech development, calibrating its opening up to the outside world and tweaking every opportunity to advance its national economic interests, persistently ignoring neo-liberal criticism and advice.

China's policy is important also because only a strong manufacturing sector can provide the basis for experimentation and innovation from which high technologies, including militarily useful technologies can later emanate.[11] This is a familiar argument made in major American reports and legislation such as the 'America COMPETES Act' and the Buy America provisions of the 'American Recovery and Reinvestment Act'. The success of this approach is displayed in not only China's larger share of every economic pie in the world compared to India, but also its innovation and social welfare indicators. Thus the Global Innovation Index developed by a team led by INSEAD ranks China at 29 and India at 62.[12]

Fourthly, an additional, and perhaps one of the Freudian reasons for closely following China's high-tech trajectory is its security dimension, as we need to understand exactly what kind of a challenge we face. With its concentration on futuristic weaponry and battle tactics, China is producing a vast array of advanced weaponry, which is upsetting the military balance in the Asia-Pacific region and causing great concern to established powers.

Finally, I have been deeply impressed by economic and technology policy strategists cum academics of Chinese origin (Prof. Justin Yifu Lin and Prof. Tai Ming Cheung). Perhaps it is because they think originally about the optimal economic/technology models that a developing country requires. In contrast, prescriptions

---

[11] There are many other essential components of a healthy national innovation ecosystem apart from the manufacturing sector that shall be tackled later.

[12] Dutta, S. (ed.), 'The Global Innovation Index 2011: Accelerating Growth and Development', 2011. Available at http://www.globalinnovationindex.org/userfiles/file/gii-2011_report.pdf.

advanced by the Washington Consensus School seem to me to offer only partial and often facile solutions to the deep structural problems that developing economies have to overcome. And after observing the effect these policies have had on a number of countries (Latin America, South-East Asia prior to the 1997 crisis and even post-USSR Russia and East Europe, and why exclude the financial collapse starting in 2008), our own policy makers should be extremely wary of swallowing their flattering exhortations.

CHINA'S MODEL

The Chinese S&T strategy comprises strong state oversight, coordination between different state organs, establishment of an S&T culture in state/private sector firms and in R&D institutions, civil military integration (CMI), educational reforms and also more traditional methods such as copying and reverse engineering. It is not possible to do justice to any of these in a brief foray such as this essay but it is attempted.

The importance of systemic reforms: There is a widespread impression that China's success is only due to reverse engineering. But China has also undertaken a large number of systemic reforms to better develop what Prof. Tai Ming Cheung categorizes as the 'hard' and 'soft' capabilities[13] required for creating a successful innovation ecosystem. As I understand it, 'hard' capabilities relate to tangibles like funding, R&D and manufacturing capabilities, number of laboratories, research institutes, etc.

Soft capabilities relate to governance, reforms, institutional synergies and efficiencies, and legal and regulatory regimes amongst other factors. Creating 'soft capabilities' is actually the 'harder' part. China has undertaken prolonged and successive reforms of its government structures, R&D infrastructure and even state-owned

---

[13] Tai Ming Cheung, 'The Chinese Defence Economy's Long March from Imitation to Innovation', *Journal of Strategic Studies*, 17 June 2011, 34:3, 325–354.

firms in order to create institutional frameworks and organs that encourage innovation.

Restructuring of industry ownership: China has permitted its defence industry firms to raise funds from the market, with 22.5 per cent 'restructuring themselves into shareholding entities by end 2007'.[14] Defence conglomerates have undertaken reforms to improve their operations, and the defence R&D apparatus has been overhauled to

> enhance basic research capabilities, diversify management oversight and funding sources from the state to the corporate sector, tear down the barriers that have kept the defence R&D system separate from the rest of the national innovation system, and forge close linkages with universities and civilian research institutes.[15]

Conglomerates have been directed to set up their own R&D centres. As a result, China is in an elite group of countries including of course the US in which business/industry funding for R&D contributes upward of 60 per cent of R&D.[16] Indeed this is a lesson for Indian industry, which has to raise its R&D contribution in national gross domestic expenditure on R&D.

Leadership: An important component of soft capabilities is leadership. In America, Government from the beginning played a major role in strengthening national S&T capacity. Government pro-activeness and not intrusiveness is essential for laying the foundations of a dynamic, indigenous, high-tech defence production sector—sitting atop a strong, diversified manufacturing sector. China has a technocratic leadership, which understands the need for technological upgradation of the economy for national security and power, and is ready to take hard decisions on the issue

---

[14] Tai Ming Cheung, 'The Chinese Defence Economy's Long March from Imitation to Innovation', *Journal of Strategic Studies*, 17 June 2011, 34:3, 325–354.

[15] Ibid.

[16] The 2012 S&E indicators state that 'the business sector, the largest performer of U.S. R&D, is also its largest funder, at about $247.4 billion in 2009 or about 62 per cent of the U.S. total'.

of re-engineering of government and prioritizing S&T through policies, mega-projects and equally mega-investments. The government is single-mindedly targeted on domestic technological upgradation that permeates all major policies. They have a focused policy framework for indigenization[17] that includes injunctions to implement CMI and reverse engineering! And China has already brought out a futuristic document recently with a time horizon of 2050! They protect domestic patents so they incentivize their own companies to innovate but of course we all know what they do with imported technologies.

Civil-military integration: China has also adopted the CMI paradigm borrowed from the US and other advanced countries. This is because the innovations produced in their huge and diversified civilian manufacturing sector, which had raced ahead of the defence economy—benefits the defence economy in a spin-on effect.[18] As Prof. Tai Ming Cheung has pointed out—and the following insights are near verbatim reproductions from his book *Fortifying China*—'as China's prosperity increased, the leadership recognized the importance of CMI. The civilian economy is constantly forced to upgrade to remain competitive while the defence economy suffers necessarily from long gestation periods. The integration of these two sectors is allowing China to leverage the more developed civilian capabilities for its military'. The CMI strategy thus synergizes the civilian and defence sectors of its economy and exploits technological discoveries in the civilian sector for military purposes, because, China understands that high-tech supply chains criss-cross both its defence and civilian sectors of its economy.

It is also helping China access advanced Western dual-use technologies through collaborations with its civilian sector that have otherwise been embargoed for China's defence economy. China

---

[17] Guidelines for the Medium- and Long-Term National Science and Technology (S&T) Development Programme (2006–2020), State Council, People's Republic of China.

[18] Tai Ming Cheung, *Fortifying China: The Struggle to Build a Modern Defense Economy*. Ithaca, NY: Cornell University Press, 2009.

has intensified its efforts to acquire high-tech companies in a wide variety of sectors abroad to aid its acquisition and assimilation of dual use technologies.

CMI is a particularly useful concept for India as it agonizes over direct and indirect offsets and a national offsets policy. There is no contradiction in permitting offsets in high-tech or educational sectors as these too form part of the ecosystem for high-tech defence production. It would be preferable if offsets resulted in investments in building up high-tech defence manufacturing in India, but investments in aviation, shipyards, telecommunications, components, high-tech alloys and composites, nano-technologies and educational institutes would benefit the overall development of the sector just as much.

Reverse engineering: The reverse engineering strategy of China is an 'APE' strategy (Assimilation, Production and Export); which is also part of their 'Guidelines for the Medium- and Long-Term National Science and Technology Development Programme—2006–2020'.[19] A reverse engineering strategy has been aggressively pursued in China and India should not shy away from following this path. 'Everybody has been there, done that', all the developed countries have done it at the earlier stages of their development and we should have no hesitation in doing so also. In fact, it is a national failure that we have not engaged in this more actively, except in the pharmaceutical sector with such good results that now they are the targets of acquisition by foreign multinationals! There are also very different stages to reverse engineering associated with the levels of sophistication that an economy has reached, starting with absolutely basic imitation, and then going on to incremental innovation. The best exposition of the different stages that China's assimilation/reverse engineering policy has graduated through

---

[19] China State Council, The official 'Guidelines for the Medium- and Long-Term National Science and Technology Development Programme' Chapter VIII state: 2. Government Official Report.

is again in Prof. Tai Ming Cheung's 'Long March...'[20] as he describes the increasingly complex forms of reverse engineering that need to be developed as an economy graduates in sophistication.

The Chinese policy, which has been pursued rather too aggressively, has naturally created a huge uproar in the West and even in Russia. The US Chamber of Commerce brought out a report in 2010 and a huge amount of pressure has been put on China to reverse these policies. Hence I do not think we have a great window or unlimited licence—excuse the pun—here. We have to be aware of the international implications a little more than China has had to contend with but on the whole reverse engineering without stealing is not something that anybody can monitor or object to.

BASIC LESSONS FOR INDIA

The lessons for India are obvious and staring us in the face.

Thinking for Ourselves: A *Desi* Consensus on National Interest

The first is the supreme importance of thinking for ourselves and abjuring intrusive external exhortations (learning from other models does not mean we do not develop our unique national solutions and question 'hand-me down solutions', and that applies to the conclusions in this chapter too!). There is a need for India to create a national model, for which a radical overhaul in the mindset as well as implementation strategy will be required, because 'Copycats don't catch mice'! The Washington Consensus has been followed so far in India with predictable results. India implemented only the 'easy reforms', with the low-hanging fruit yielding immediate efficiency gains and growth. But the reforms, while having many

---

[20] Tai Ming Cheung, 'The Chinese Defence Economy's Long March from Imitation to Innovation', *Journal of Strategic Studies*, 17 June 2011, 34:3.

positive results, also yielded lop-sided results, such as jobless growth, a stagnant manufacturing sector in terms of increasing share in GDP and R&D deficits. Sure enough, as earlier pointed out, this led to a huge deficit on the trade account and the sharp devaluation of the rupee. As pointed out by me in 2001, 'Neglecting infrastructure, India did not derive full advantage from the limited liberalization effected in this phase (1991–1993). The current economic slowdown and whittling down of the manufacturing sector in India can be attributed to the neglect of core sector reforms, something China tackled early on, with a host of repercussions for sustained growth in many sectors'.[21] So this has happened before and will happen again unless we intervene to break this vicious cycle.

Leadership

The second issue is that of leadership. Successful reform depends on clearing the impediments in the way and negotiating bureaucratic mazes, requiring the kind of over-arching direction that only the Prime Minister's office can provide.

Thus at the first meeting of the High-Tech Defence Innovation Forum,[22] we produced a 'Vision for an Integrated Science & Technology Advancement Strategy', appropriately nicknamed VISTAS, a tad better than the American acronym COMPETES 'Creating Opportunities to Meaningfully Promote Excellence in Technology, Education, and Science'. Leadership was one of the first issues we tackled. We proposed establishing an apex body—the National Technology Advisory Council (NTAC)—representing all stakeholders from industry, academia, government and the armed

---

[21] 'Can India Overtake China?' by Smita Purushottam, Fellow, Harvard University, 2000–2001.

[22] The Forum meeting was organized at IDSA on 14 July 2011, to formulate a strategy for indigenous high-tech defence production, bringing together academics, military officers, scientists, experts and senior bureaucrats.

forces, directly under the Prime Minister. All institutions and departments concerned with technology and defence would report to the NTAC. It appears that the US has created the President's Council of Advisors on S&T to maintain US leadership in S&T, as it recognizes that technology has been the key to US power and dominance.

Strengthen Overall Manufacturing, Infrastructure and R&D

It must be understood that it is foolhardy to attempt establishing a high-tech defence sector in isolation from the rest of the economy. China has understood very well that only a healthy and diversified manufacturing sector can support an advanced defence industrial sector and its 16-character policy outlining the principles of CMI encapsulates this philosophy that a sophisticated defence economy cannot thrive without a sophisticated economy-wide supply chain. The third basic lesson is therefore to make the promotion of overall manufacturing, infrastructure and indigenous R&D in the defence and civilian economies. A more aggressive, nationalistic pursuit of manufacturing prowess is therefore required. Basic principles that one would like to see in a government policy include a commitment to strengthen India's existing manufacturing sector and implement an 'India First' policy across the board in offsets, manufacturing, transfer of technology, FDI, tariff policy, etc. If America can have 'Buy America provisions' in their stimulus bill, why cannot a developing country like India have 'Buy Indian' provisions? In fact, this is present in the defence offsets policy but has to be implemented more aggressively in every sector.

In our strategy, we should:

Strengthen, not weaken, and expand India's manufacturing sector: This means imparting greater dynamism to economic reform and strengthen the manufacturing sector, because India cannot develop cutting-edge technology without a diversified manufacturing base.

Cutting down unnecessary bureaucratic hurdles, improving India's ranking in the ease of Doing Business Index from an abysmal 132 and tackling governance and infrastructure issues are demands that have been made and need to be implemented forthwith. The processes for setting up businesses must be drastically reduced and simplified. Furthermore,

- We should retain 'Buy India' provisions at least in the government sector.
- Foreign takeovers of our thriving pharmaceutical companies should not be permitted. A finely calibrated policy to encourage domestic R&D must be formulated instead, while permitting our companies to reap the benefits of reverse engineering. Giving up our companies to takeovers will *not* produce domestic R&D.
- We should protect the interest of our manufacturing sector under free trade agreements including the interests of our automotive sector under the India–EU FTA. Again, policy has to ensure that the sector continues to upgrade its products and invest in R&D.
- Local sourcing obligations for FDI in retail should be set at 80 per cent. The protests of foreign investors against local sourcing obligations, whether these are in retail or under offsets, should be dealt with firmly. There is no case for timidity on this front. This will give a major fillip to domestic manufacturing.

Leverage Defence Reforms: Declare a Defence Industrialization Strategy as follows: Hundred per cent offset obligations under defence offsets need to be set. Enough studies have been done to show that the majority of countries have 100 per cent and more defence offsets. This can generate a massive re-industrialization wave. In offsets policies, Indian industry must be prioritized in all defence and civilian acquisitions and Indian firms or consortia must be designated as lead integrators for defence acquisition and all other high-tech and labour-intensive manufacturing projects.

While 30 per cent can continue to be allotted to the defence production sector, 70 per cent can be used in high-tech sectors such as aerospace, telecommunications, railways, composites, machine tools, electronics hardware and other requirements of India's hightech industry.

Furthermore, we should go for CMI and do away with the bickering on the merits of direct and indirect offsets—both are equally useful if strategically applied. This would create a dual use manufacturing base benefiting the defence sector, the philosophy behind CMI in China.

Lastly, the National Offsets Policy should be announced forthwith with 50–100 per cent offset obligations. Offset agencies in all the other sectors depending on huge imports such as railways, telecom, aerospace and IT should be set up and the proceeds used to launch a renewed manufacturing revolution.

Increase R&D funding and encourage the R&D culture everywhere: We should go for reverse engineering on a war footing. China has not suffered consequences of its aggressive reverse engineering policies. R&D funding must increase across government, industry and services. China has announced it will commit $1.5 trillion for R&D expenditures under the Strategic Emerging Industries Decision. This seems to be in addition to the investments in R&D already made that has catapulted China to the position as the world's second largest spender on R&D as noted at the beginning. We must realize that no one will give state-of-the-art technology to you; you have to develop it yourself. Indeed, India's R&D expenditure should be doubled to 1.75 per cent of GDP by 2015 and 2.5 per cent by 2025. Likewise, industry should contribute an increasing proportion of R&D expenditure. We should remember the international example where industry is in fact contributing a greater proportion of national R&D expenditure. We can further incentivize the private sector to invest in innovation while making it mandatory. The service sector especially in telecommunication and aerospace should invest in indigenous technology and manufacturing capabilities.

Restructure Defence Public Sector Undertakings (DPSUs), research and educational infrastructure:

1. Modern management must be introduced in PSUs, including DPSUs and research institutes to make them more effective and result-oriented. They should be liberated from the culture of patronage.
2. The best practices of MNC R&D centres in India should be studied. If the Indians in the MNC R&D centres can produce ground-breaking technology, they can do it for local industries too.
3. We need to reform and massively increase investment in the education sector. China has launched several projects to restructure, reform and expand its higher educational structure and facilities, and has become a magnet for international students. India can easily outclass China with the right policies. Defence offsets can be used to fund a massive expansion of higher and vocational education in India.

CONCLUSION

India has to overcome its hesitations and delays in launching fresh reforms, which include a high-tech defence industrialization and innovation drive if it wishes to catch up with the world's technological leaders. Achievement of our goals will help meet our requirements for both high-tech and labour-intensive products, thus reducing our import bills, providing employment and guaranteeing greater long-term national security. Without the aforementioned reforms and changes, India will neither become an advanced technological nation, nor will it succeed in uplifting the welfare of its citizens. And if the West genuinely wishes Indian democracy to flourish, it will support the path discussed in this chapter.

7

Technology in Offsets: A Strategic Tool to Galvanize the Indian Defence Industry

Shobhana Joshi

Abstract

The primary aim of the offset policy of the Ministry of Defence has been to leverage the massive investment for defence acquisition to strengthen and energize the defence industrial base of the country. Defence procurements from the foreign vendors under the offset umbrella have long been viewed as the middle path between acquisition of defence equipment through imports and indigenous R&D and production, which will also provide benefits to the domestic industry. This chapter analyses whether transfer of technology through offsets can prove to be an effective method of technology acquisition and reduce the complexities between the buyer and the seller. It examines the study presented to the US Congress on various facets of offset implementation and its impact on industrial activity and employment. Various approaches to valuation of technology have been explained. The chapter focuses on the success story of South Korea's offset programme and finally provides an insight, based on a personal perception, on a range of policy prescriptions like buyer-directed offsets, a mandatory cap for defence products, a higher percentage towards core technologies, consolidation in offset volumes, third-party eligibility, etc., by which India can take the benefit of technology in its offsets and raise the capability bar of its domestic defence industry.

INTRODUCTION

India is a recent entrant to the group of 130 countries that incorporate offsets in their defence contracts, which are valued at US$5 billion each year and range between 5 per cent and 30 per cent of the global trade. The offset policy in India was introduced in 2005 as a part of the Defence Procurement Procedure. This is not surprising, considering that the most noticeable increase has taken place in the capital acquisition head from ₹26,926 crore ($4.80 billion) in 2005–2006 to ₹75,148.03 ($12.52 billion) in 2014–2015 (budget estimates). The primary aim of the offset policy of the Ministry of Defence since its inception has been to utilize this massive investment in defence modernization to provide a fillip to the Indian defence industry both in the public and private sectors. Though from time to time certain groups have voiced the view that the offset programme in India needs to be relaxed to include other sectors, from the government's perspective it is very clear that offsets are primarily to strengthen the defence industrial base in the country.

OFFSETS: DIRECT/INDIRECT AND THE INDIAN INTERPRETATION

Basically, offsets are classified into direct and indirect offsets; direct offsets are determined by the relationship of the offset arrangements to the subject matter of the procurement that is, it is related to the contract and requires manufacture of components that are linked to the equipment contracted; for example, if you are buying an aircraft then the accompanying manufacturing or technology for aerospace should be provided and not for armoured vehicles or a naval ship. An indirect offset is when there is no relation between the subject matter of the sales agreement and the components being manufactured. Now if you go by this classical definition of direct and indirect offsets, then India's offset policy cannot be considered direct because although it envisages that the offset obligations should be related to the defence sector, it is not necessarily linked to the weapon system or equipment for which the contract has been signed. This nuance in the policy is important to understand the nature of direct offsets in the Indian context. However, at the same time, this does not take the edge off the policy but makes it less rigid in implementation so long as the benefits are passed on to the defence sector. The range of offset transactions being implemented by various countries includes sub-contracts, co-production, overseas investment, technology transfer, licence production, counter-trade, miscellaneous and purchases. However, what needs to be noted is that some of them are common to both. Technology transfer can either be a direct offset or an indirect offset also.

WHY OFFSETS AS AN OPTION

When we view the various methods of defence acquisition in relation to the degree of industrial involvement, at one end of the spectrum is off-the-shelf import of equipment, which does not involve any development or production for the domestic sector and at the other end is indigenous development and production of equipment where the entire work is allocated to the domestic defence industry. Somewhere in between is defence procurement from the foreign vendor under the offset umbrella, which is linked to industrial benefit for the domestic industry. However, from a short-term financial perspective normally off-the-shelf equipment would be the cheapest option because there is no investment either in the research and development (R&D) or in the production. Indigenous development naturally would have implications, both in terms of cost and time as the government has to take the entire responsibility for the R&D. That is the reason why most countries have found the via media of using the offset option for getting the benefits for the domestic industry.

OFFSETS AS DRIVERS OF TECHNOLOGY

The reason why offsets are viewed as a driver for technology is that when technology is sought to be purchased, critical technologies are denied and there are various control regimes under which you are supposed to sign a number of agreements such as International Traffic in Arms Regulations (ITAR), etc. In addition, the exorbitant pricing of the technology modules also becomes a prohibitive factor raising questions of financial viability. Transfer of technology also has several preconditions like the numbers to be manufactured, restrictions on modifications and upgrades and so on. There are global success stories such as Israel, South Korea and China that demonstrate the effective use of offsets for upgrading the domestic industrial capability. In fact, Israel and South Korea have become arms exporting countries. The defining aspects of the policies of these countries have been the focus on the acquisition of high technology with priority to military technology. Both are now amongst the top five largest offset beneficiaries from US companies.

HOW OFFSET OBJECTIVES EVOLVE

The trend in the evolution of offset objectives, not only in the context of India, but in other countries as well, is the shift of emphasis from general economic development to domestic industrial production capacity in specific core technologies. The initial objective of the policy would be to enhance the overall economic development goal; one would like an offset policy to be a facilitator to build roads, to build schools and so on. That may be the objective of most of the developing countries. Later on, the focus shifts to technology transfer tied to the local industry. Countries also try to avoid offset outsourcing of low-technology work and look for new and sustainable work to enhance the technological capability that is not available in the domestic industry. That point is very valid because we do not want offset to just produce what we are already producing in the country. The idea is to use offset to obtain what we do not have. And as the policy evolves, we also learn from experience and get wiser. Some countries also target to achieve niche capability in certain strategic technologies and use offsets to obtain such technologies.

OVERVIEW OF OFFSET IMPLEMENTATION

Will technology through the offset route be the game changer and transform the domestic defence industry? A reality check needs to be done to ascertain what has actually been achieved by countries that have incorporated transfer of technology in their policy. The single-most comprehensive document on offsets that is available is the study carried out by the Bureau of Industry and Security of the US government, which is presented to the Congress every year. The 16th Report indicates that during the 18-year period from 1993 to 2010, the top three offset transaction by categories are purchase, sub-contracting and technology transfer under both direct and indirect offsets. However, in the direct offsets category over 50 per cent is by sub-contracting. It is significant to note that though the US government provides transfer of technology to a number of countries, technology transfer accounts for only 18.59 per cent of the total offset percentages, the purchase category is highest, at about 36.69 per cent followed by sub-contracting at 21.32 per cent in terms of actual value. This is basically just to highlight that defence technology is a very complex area and even though there may be numerous countries having the technological base to absorb such technologies yet the manufacturers prefer not to part with technology and to take the easier option. For them purchase or manufacturing would be the much easier option than giving the technology. Another important statistic that this report has highlighted was that by industry sector, during 2009–2010, 71.5 per cent of reported offset transactions were related to manufacturing.

In manufacturing, the top four industry sectors were (1) aircraft manufacturing, (2) aircraft parts and auxiliary equipment, (3) aircraft engine and engine parts, and (4) search, detection, navigation, guidance, aeronautical, nautical systems and instrument manufacturing. Together these four sectors accounted for about 41 per cent of the manufacturing sector.

IMPACT ON INDUSTRIAL BASE AND EMPLOYMENT

As a country that has only recently incorporated the offset provision in order to leverage its buying power as part of its defence acquisition, an understanding of how the offset providing countries view offsets vis-à-vis its impact on their economy is necessary. Therefore, while the BIS study identifies the potential economic and industrial base benefits like lower overhead and unit costs, interoperability of defence systems, defence export/sales, as an important component of revenue generation for their manufacturers, they bring out the negative impact on US industrial activity and employment as well. What has been emphasized is that offset activity displaces work from the US and that technology transfer may increase R&D spending in foreign countries and capital investment creating competition to the US industry. The impact analysis for 2010 puts the total value of inputs for defence exports as $12.33 billion, which would create or sustain about 45,576 jobs. On the other hand, the negative economic activities are the jobs that were taken away which actually resulted in a net reduction of 23,022 jobs. Though the overall effect still resulted in a net gain of 22,553 jobs arising from export sales contracts with associated offset agreements, typically the US considers that any economic activity denied in terms of gain for manufacturing opportunities in their country, is a loss. Therefore, from such an impact analysis one can surmise that there would be a value added because of obtaining offsets, in other words, there would be a price for offsets.

OFFSETS AS A ROUTE FOR TECHNOLOGY ACQUISITION

Technology transfer through offsets is an efficient method of technology acquisition because in the direct purchase of technology, the entire risk is on the buyer but when technology is part of offsets then the risk shifts to the seller as well as he has greater incentive to transfer the technology as it is linked to the acquisition of the entire system. However, the BIS report does not collect data on specific technologies transferred but based on anecdotal information obtained from industry it has been stated that cutting-edge or nascent technologies under development in the US are less likely to be transferred to foreign companies than more mature technologies. Therefore, even through the offset route frontline and cutting-edge technologies may continue to be denied.

VALUATION OF TECHNOLOGY: HOW TO UNRAVEL THE CONUNDRUM

According to analysts, as offset trade is negotiated in terms of valuation, various methodologies are used to bridge the gap between the perspective of the buyer and the seller. The buyer focuses on the value because offsets provide an opportunity to get the desired defence technology. For the seller, on the other hand, the cost incurred to implement the offsets assumes significance. The three standard approaches that are used for evaluation of technologies are cost that is based on the cost to create it; market that is based on the comparable market price on similar technology—transfer transactions; and income that is based on economic benefits derived from owning the technology. The various elements that define the valuation process could range from the investment made in the R&D by the sellers company/government, the market value, job losses, the amount needed for domestic development, anticipated revenues, jobs created in buyer's country and benefit to local economy, the cost of jigs, tooling, etc. The trade-offs between the

buyer and the seller is crucial and is evident in numerous negotiations in which I have participated where the seller tries to get
maximum compensation for its technology investment and the
negative economic impact and the buyer looks for cost-effective
terms for the transfer of technology. For the buyer or the recipient,
the bottom line is that the price of the equipment should be reasonable after the cost of the technology is amortized on the minimum
order quantities. But if it does not bring down the indigenous price
after factoring in the time taken to absorb the technology and the
learning curve then acquiring that technology may not be financially viable and it would need to be a strategic decision. These
are the kinds of issues which one goes through while evaluating
technology options. There is no single one-size-fits-all formula and
it has to be a mix of various kinds of parameters. Two research-
ers from South Korea noted that most buyer countries do not
possess the means of estimating the value of defence offsets due
lack of an appropriate valuation model. Based on their research,
they have developed a Defence Offset Valuation model that is a
combination of various methods like cost, income, lines of code
and case studies methods model in order to increase objectivity
and credibility.

MULTIPLIERS AS INCENTIVE FOR TARGET AREAS

There is always a value gap between the offset provider and the
recipient. The recipient countries use multipliers as an incentive for targeted areas and these multipliers work as leverage to
bridge the value gaps between the offset provider and the recipient. However, in the case of offsets, the technology itself is not the
ultimate purpose of procurement; it is the weapon system. Korea
is an example of an effective and strong offsets policy with technology transfer as a cornerstone of its policy. As per the economic
indicators, the defence exports of South Korea were $1.1 billion in
2009 and the offset trade ratio was 42 per cent, showing a remark-
able increase from 6.3 per cent in 1998. Some of the key policy

parameters that have contributed to the success of South Korea's offset programme are:

- Stringent value addition requirements in South Korea.
- At least 40 per cent of domestic manufacturing to directly relate to weapon system being procured.
- Government holds ownership rights on all intellectual property rights (IPRs).

OFFSET POLICY: NEW HORIZONS

Basically, what are the policy prescriptions by which India can take the benefit of technology in its offsets? Here, I would like to point out that the areas highlighted are not necessarily the views of the government but based on a personal perception after having dealt with the sector for a long time and been part of the changes which have taken place in defence acquisition and procurement procedures and practices. In any case, all policy is arrived at through a process of consultation and consensus.

First, the defence offset policy is now at a stage when the focus needs to shift to the requirements of the buyer and the kind of defence manufacturing base it is aspiring to achieve. The objective of the policy should not be merely for the Indian industry to become part of the global supply chain but to have capability for design, integration and manufacture of high technology and sophisticated weapon systems from smart munitions, electronic warfare systems, night vision devices to ships, submarines, tanks and aircrafts. The offsets need to be directed to obtain the specific technologies to enable the country to reach this goal.

Second, there is a need to fix higher percentage towards core technology transfer for defence vis-à-vis non-defence namely civilian aerospace and homeland security sectors, at say 70:30 with 70 for defence and 30 for non-defence. Otherwise we will get technology only for civilian aerospace and homeland security and may not actually get the desired technology needed to fill the gap in critical technologies in the country for the defence sector.

Third, there should be greater emphasis on manufacture of defence products for exports to niche markets with strategic value and therefore a 25 per cent mandatory cap for defence products needs to be introduced.

Fourth, third-party eligibility should not be diluted beyond the prime contractor—we have gone on to sub-contractor—but there should be no further relaxation.

Finally, there is a need to move towards consolidation in offsets to get the advantage of volumes, and therefore the list of potential Indian offset partners (IOPs) in the offset proposals should be limited to not more than four and changes allowed only within these. The list of offset components should also not exceed a specified number.

CONCLUSION

Even during this short span, India's offset policy has succeeded in energizing the domestic industry. Twenty-four offset contracts valued at ₹28,770 crore ($4.87 billion) have so far been signed. Another 44 cases are in the pipeline, which will take the quantum of offset obligations to approximately $15–16 billion. The country is therefore poised to move to the next level and technology-driven offsets will be the trajectory that raises the capability bar of the domestic defence industry.

BIBLIOGRAPHY

Defence Procurement Procedure 2013 (incorporating the Defence Offset Policy), Ministry of Defence, New Delhi.

Jang, W.J. and Joung, T.Y. 'The Defence Offset Valuation Model', *DISAM Journal*, December 2007.

Lee, C.J., Jang, W.J. and Yoon, B.K. 'Technology Acquisition Policy and Value Efficiency Analysis on Offsets in Defence Trade', *DISAM Journal*, 2011.

Radhakrishnan, P.S. and Kumar, P. 'Leveraging Offsets for Self Reliance in Defence Technologies', *Journal of Defence Studies*, January 2009.

Ravindran, S.P. 'Technology Inflows: Issues, Challenges and Methodology', *Journal of Defence Studies*, January 2009.

US Department of Commerce, Bureau of Industry and Security, Offsets in Defence Trade: Sixteenth Report to Congress, 2012.

Verman, S. 'Constructing an Effective Offsets Programme: Drawing Lessons from the South Korean Experience', *The Administrator*, Vol. 52, No. 1, 2012.

8

The Value Chain in Defence Offsets

K.V. Kuber

Abstract

Offsets will make a difference to the industry if the industry makes business sense of the transactions involved. There are numerous examples across the world of companies either having done well and going forward or having sustained losses in the bargain. The Indian offset policy is riddled with interpretations due to the fuzzy nature of policy formulation. There are pieces of extant regulations, which need to be looked for, mostly from different ministries, such as the Ministry of Commerce and the Ministry of Finance. The key to success of offsets lies in generating a sound business plan, getting into the global supply chain of the original equipment manufacturers, cutting costs and producing an efficient plan. Offsets cannot be a business model by itself; it needs to be integrated into the overall business strategy of the company. For any offsets to succeed, the business plan must not be generated or reviewed from an offsets perspective. However, the converse is true: a good business plan can incidentally be translated into an offset proposal, which will always work.

INTRODUCTION

Offsets policy is a magic wand: swing it, and with it everything will change; we create a dynamic defence industry and everything will work out; so is the belief, reality is quite different.

Offsets have been introduced in the defence procurements with an objective of expanding and strengthening the defence industrial base. The one fundamental positive aspect of offsets is that it provides a guaranteed business to the domestic industry, by making it obligatory for the foreign OEMs desirous of participating in defence contracts to engage with the Indian industry, in terms of defined industrial participation. The guidelines for such participation are laid down in the form of the, 'Procedure for Implementing Offsets', popularly referred to as the offsets policy.

The Indian experience with structured defence offsets commenced in 2006 through the introduction of offset obligations in the Defence Procurement Procedures (DPP). These were then subsequently revised in DPP 2008 and DPP 2013 as part of an ongoing process. To understand defence offsets and their context in the recent past, it is important to recognize, note and realize the factors that impact this policy.

The primary factor that influences the offset policy is a strong need for self-reliance, probably articulated for the first time as part of the DPP. It is critical that the country provides to the armed forces, the capability to sustain long-drawn conflicts with complete support related to serviceability/availability of equipment or supplies. This is interlinked to the current dependence on exports and the need to inverse the trend to create an Indian defence manufacturing industry that can sustain the needs of our armed forces.

Linked to the above is the intent of the government to leverage its buying power and invite global OEMs to provide market access for products manufactured in India thereby providing a jump-start to businesses that can comply with the stringent defence manufacturing requirements. This has had a great influence in the formulation of 'Buy Indian' category and the articulation later revealed in the Defence Production Policy.

Offsets provide Indian industry with a catalyst to quick-start the defence manufacturing phase. It is by no means a provision for inefficient manufacturing to be sustained through a popular government policy. Offsets were meant to be an opportunity for Indian industry to utilize its traditional advantages of large pool of talent, large domestic demand, significant cost benefit in input costs, a strong legal framework to infuse and sustain technology to world standards, and secure a pie of the lucrative global defence business. The opportunity perspective for government and private industry is given in Figure 8.1.

DEFENCE OFFSETS IN INDIA: THE CURRENT SCENARIO

The applicability of offset clause is as listed below:

1. Under the current DPPs, the offset provisions are applicable to all capital acquisitions categorized as 'Buy (Global)', that is, outright purchase from foreign/Indian vendor, or 'Buy and Make with Transfer of Technology (ToT)', that is, purchase from foreign vendor followed by licensed production, where the estimated cost of the acquisition proposal is ₹300 crore or more.
2. The DPP also prescribes a minimum offset obligation of 30 per cent of the estimated cost of the acquisition in capital acquisition programmes where offset is applicable. This value of 30 per cent is only the lower benchmark

Figure 8.1　Offset opportunity perspective

Offset Opportunity Perspective

Government Perspective

Leverage buying power to create a defence manufacturing base that addresses the need of self-reliance

Private Industry Perspective

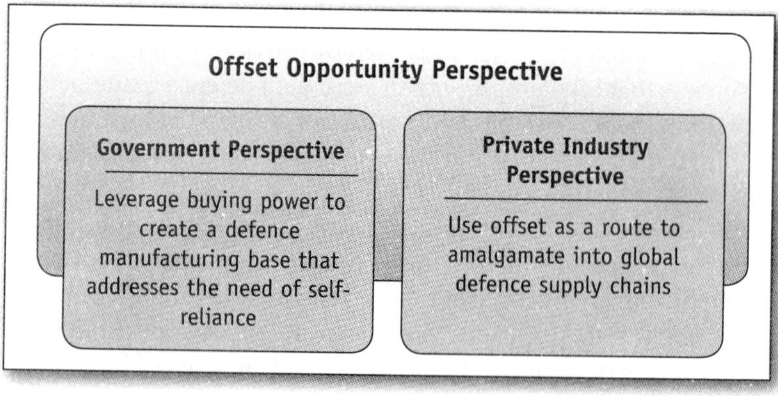

Use offset as a route to amalgamate into global defence supply chains

and the government reserves the right to have higher offset percentages for some capital acquisitions [case in point being the medium multi-role combat aircraft (MMRCA) contract where an offset obligation of 50 per cent is stipulated].

It is worth understanding as to what constitutes offsets?

1. India follows a unique model of what constitutes offset wherein global OEMs and international vendors are not restricted to the platform they are supplying and can support the broad-based development of Indian defence industry through sourcing of defence products or services from across the entire defence sector. A list of what constitutes a defence product for offset is provided in the DPP 2008. This has since been expanded to include synergistic sectors like civil aerospace and internal security (more popularly known as homeland security globally). The entire concept of 'defence product' has been replaced by 'eligible product' since DPP 2011. This is a welcome change for the larger industry that now has a wider choice for participating with OEMs. India has also increased its budget for internal security with huge allocations for paramilitary forces, police and many such institutions, for safe cities projects and many such projects taken up by a number of states.

2. The Indian understanding of offset is a practical middle path of what is generally constituted in the west as direct offsets (i.e. platform-linked offsets) or indirect offsets (i.e. non-platform-linked and non-military-oriented offsets). This has been further expanded and we also have indications for inclusion of the commercial shipbuilding within the scope of offsets.

3. A shot at the implications of the newly announced policy is given at the end.

The current policy stipulates six methods for OEMs and foreign vendors to discharge Offset obligations. These six methods can be used in any combination to discharge. These include:

1. Direct purchase of eligible (defence, inland security, coastal security and civil aerospace) products and components manufactured by or services provided by Indian enterprises.
2. Purchase of services (maintenance, repair, overhaul, upgradation, life extension, engineering, design, testing of eligible products, software development, quality assurance and training) from Indian industries.
3. Execution of export orders by Indian industries. Execution of export order implies securing a specific contract for Indian industries for any of the above two ways (i.e. purchase of defence products and components or services) from any other global entity.
4. Foreign direct investment (FDI) in joint ventures with Indian enterprises (equity investment) for manufacture or maintenance of eligible products and components.
5. Investment in kind in terms of ToT, to Indian organizations engaged in manufacture or services. The Ministry of Defence (MoD) will also give a multiplier of 1.1 for such ToT, on the value of buy-back.
6. Investment in kind in terms of equipment through non-equity route to Indian enterprises.
7. Provision of equipment and or ToT to government enterprises.
8. Technology acquisition by DRDO with multipliers up to 300 per cent.

The policy is dramatically different from the earlier versions, in that it includes most concerns voiced by the OEMs as well as the Indian industry. The best practices of the larger group practicing offsets across the world have been taken into account. The scope of discharge of offset obligations has been expanded and more avenues are available in terms of ToT, etc., OEM as well as a larger industry segment can participate. Micro-small-and-medium-sized enterprise (MSME) segment has been incentivized, by the choice of which, the OEM will get an additional multiplier of 1.5. This means, when an OEM chooses an MSME as Indian offset partner (IOP), then for a trade value of say US$10 million, the corresponding offset credit accrued to the OEM is US$15 million. This is a great incentive to choose MSMEs.

The validity of banked credits has been extended and risks of penalty mitigated to a great extent. The management mechanism has been more clearly defined. These are the positives of the new policy. The policy encourages manufacturing technology to be transferred, with a small multiplier of 1.1, to incentivize the OEM and a larger multiplier up to 3.0 for technology acquisitions. It allows for investment in kind, both through the equity route as well as the non-equity route, coupled with a compulsory buy-back by the OEM. A minimum of 40 per cent buy-back is mandated, thus maintaining checks and balances. It is a progressive policy, with adequate safe guards for the incentives provided, to have a regulatory oversight mechanism, on the offset transactions. Risk mitigation has been attempted to keep the cost of acquisition low by capping the penalties at a maximum of 20 per cent for non-compliance of offset schedule. At the same time, the OEMs have been incentivized, by allowing an additional 24 months for discharge of offset obligations.

The policy has maintained the original stance of value addition in India, and has further clarified the position with regard to the non-allowable content. This is a welcome step, although the direction is unclear. Maybe, it could have catered for an incentive for value addition to encourage inclusive growth. Fresh entrants may face some difficulties here and OEMs may resort to the traditional players in this segment. A diagrammatic representation of the methods of discharge of offsets is given in Figure 8.2.

THE RECIPIENTS OF OFFSETS

It is critical to note that the current policy has gone far beyond the initial definition of Indian defence industry to now 'Indian

Figure 8.2 Methods of discharge of offsets

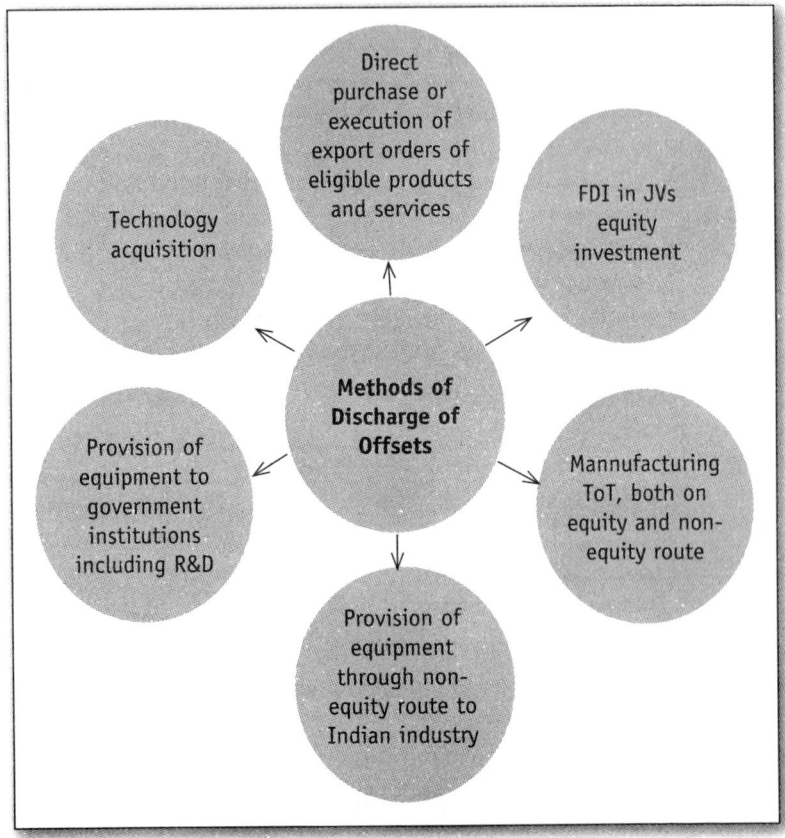

Enterprises'. This includes Defence Public Sector Undertakings (DPSUs), the Ordnance Factory Board (OFB) and private Indian industry, internal security and civil aerospace sectors, government institutions and establishments. It is also important to note that through other significant regulations private Indian industry is mandated to comply with the following, in order to qualify as being a valid Indian Offset Partner:

1. Owned up to a minimum of 51 per cent by resident Indian businesses (hence up to 49 per cent by the foreign OEM).

2. Controlled (through a majority of directorships) by Indians.
3. Incorporated in India.

It may also be noted that Indian companies need to comply with Department of Industrial Policy and Promotion (DIPP) guidelines for licence requirement for undertaking defence manufacturing, only where it is applicable. This actually means that the need for a compulsory licensing has been done away with and only those products that are otherwise included in the restrictive list will need an industrial licence (IL) from the DIPP. Fully owned subsidiaries could also be eligible for discharge of offsets if they are compliant with the relevant guidelines of the DIPP.

TECHNOLOGY AND OFFSETS

Technology will never come free. This has been the experience worldwide as well as domestically. Foreign OEMs always try to find out how they can discharge offsets without having to part with technology. Even today in the advanced world and in the most technologically advanced country, the US, the DARPA funds the entire technological initiatives. Every research that goes into any of the products or the future innovations is something that is funded by the government, owned by the government. The government has to come forward to take initiatives in funding technology and Defence Research & Development Organisation (DRDO) is our best bet in this regards. It may be worthwhile for the government to find methods to fund technologically sensitive projects with the private entities. The private sector by itself cannot fund research and development (R&D) when they are not assured of returns. The DPP still stipulates the concept of developing a multi-vendor situation and evaluation that leads to the lowest bidder, most popularly known as L1. So, we should not expect the private sector to keep investing in R&D and then the MoD as always would make acquisitions on a multi-vendor basis. This is clearly not in the interest of indigenous design and development.

MoD is seized of this requirement of technology transfers and consequently has accepted a model worked out by DRDO for inclusion of ToT as a valid method of discharge. Associated with this are the entire issues related to FDI and the restrictive cap of 26 per cent in the defence sector. This has now been relaxed to 49 per cent. This is expected to be a game changer, although the government has made it amply clear that 49 per cent is not written in stone. Higher percentages of offsets will be allowed, if there is a merit in the technology being offered for transfer and if it will create a real value in the country.

The previous versions of the policy on offsets did not include technology transfers as a valid method of discharge of offsets obligations. This is because the valuation models were not in place. The MoD has now, at the behest of DRDO, come out with technology transfers of two types. The first one is related to the investment in kind in terms of ToT to Indian enterprises, either through joint ventures (JVs) or through the non-equity route for co-production, co-development and production or licensed production of eligible products and services. The MoD expects the OEM to part with ToT in full, without charging any licence fee and deliver the complete documentation, training and consultancy required, without any restriction on domestic production, sale or export. In such a case the government will accept such ToT as a valid offset discharge and also provide an incentive of a multiplier of 1.1 for any buy-back by the OEM on the product that is the result of such ToT. Therefore, in principle, manufacturing ToT is allowed by the MoD, with suitable incentives. This has been a long pending demand of many foreign OEMs, who believe that the manufacture of the required product could not be affected, due to lack of such capability in the domestic industry and they had to perforce transfer some amount of manufacturing technology in terms of drawings and special designs, for a 'build to specs' and 'design to build' manufacture. This is a welcome step and the industry would benefit from this.

The second one is in terms of technology acquisition, by the DRDO. For the first time, the MoD has provided an indicative list of technologies they are on the lookout for. A detailed procedure for the technology acquisition process has been promulgated and the OEMs will have a close look at this for making any proposals in this area. This may be more difficult to come by, since technology is pretty close to anyone who has developed it and they may find the going pretty tough to be able to part with that. Therefore, we could see many flavours of vintage technology being offered and since fortunately DRDO is playing a central role in making assessments and recommending acceptance, we expect to witness some very intense deliberations on every proposal that comes by.

It is, however, a fact that Indian industry desperately needs technology to be able to come up to global manufacturing standards. It many times appears that it is no one's responsibility to equip the armed forces with the state of the art equipment and systems to enable them to discharge the responsibility of national security placed upon them by this nation. Technology acquisition through DRDO will be a step in the right direction, since DRDO would absorb the technology being offered (so OEMs cannot complain that the Indian industry is not yet ready to absorb technology) and then transfer the same to the Indian industry for manufacture. Domestic industry will actually gain from such a transaction.

Manufacturing technology is likely to be made available through the route of offsets, if it does not come in the way of loss of opportunity. In this competitive environment prevalent across the globe, where each nation is striving to protect and promote their own economy through export oriented revenue generation; to expect transfer of home-grown technologies at throw-away prices through offsets is probably wishful thinking. Even in the case of manufacturing technologies, this is better achievable through the route of JVs, when both partners can visualize a definite business plan.

MANUFACTURING INDUSTRY AND OFFSETS

Manufacturing of defence goods continues to be governed by the requirements of licensing. The government has tried to find a way out of this requirement though a satisfactory solution is still a long way off. The requirement of licensing has been an impediment for many companies who intend to make a foray into manufacturing. This is essentially because of the fuzzy policies and lack of clarity. Most of the time, companies are running from pillar to post trying to figure out if they indeed need a manufacturing licence and when the requirement is felt, they wait for more than 18 months to be told whether they get one or not. Lack of communication and transparency, from the MoD licensing processing authorities is a major source of irritation.

For the Indian industry, leveraging the offset opportunity is further complicated by the stringent requirements of export control through SCOMET. It is understandable that world-wide export control does play a significant role; however for a nascent and developing defence manufacturing industry like India's—a simpler procedure will go a long way in ensuring efficient implementation of the policy. Many companies in this space, who have obtained an IL for a defence product, wait for more than 8 to 10 months to get an export clearance order. While one of the stated objectives of Defence Offset Facilitation Agency (DOFA) is, 'to promote exports of defence goods and services', DOFA in many cases has been an impediment for the same. The case is even worse, due to the fact that many companies, who had gone the full nine yards in obtaining an export clearance, have to wait for similar periods again even when they need to export the same item once again. This pain has to be endured by companies each time they intend to export, clearly indicative of the fact that the government is indeed not promoting and is far from it. There is another crucial aspect, the Chairman of the DOFA redesignated by MoD as JS (Defence Offset Monitoring Wing, DOMW), will be the same personality who will also chair the committee that recommends issue of IL to the DIPP. There must be a different agency for this, since both functions, one being the cause for the other should not be chaired by the same official. Human biases work very strongly, either in favour or against a particular company, and we also need to keep in mind that since complete authority is concentrated in one person, little room is left for professional discussions.

In DPP 2011, a listing of the defence goods, export of which counts towards liquidation of offset obligation, was provided. This list, however, has elements that are not a part of the specific chapters dedicated to defence and aerospace manufacturing as per International Trade Classification, harmonized Code (ITC-HS). This creates confusion for the manufacturer. A rationalization of the defence-related products in the ITC-HS will help in enabling the manufacturer to have a clearer picture. The MoD must clearly state their requirements for licensing and not take the support of extant regulations and expect the industry to run around in circles to find out the relevant regulations.

DIRECT PURCHASE OF DEFENCE GOODS

Taxation in India plays a very dominant role in sourcing supplies from India. On an average, tax liabilities could be as high as 36–40 per cent dependent on the state or region of sourcing. As the OEMs can source products (direct purchase of eligible products) from IOP, if the said product is delivered to a destination within the country, then all taxes as applicable are liable, thus escalating the cost. This has to be borne in mind, both for the foreign OEM and the IOP.

Under the current regime, a peculiar situation emerges if a foreign vendor identifies a partner to do assembly of a system in India (generally called integration). While integration is a valid activity under a programme, the products sourced or supplied attract heavy taxes and duties. This peculiarity also covers the situation where a production agency is nominated for a particular programme. Supplies made by private Indian industry in specific cases where the lead integrator is working on a deemed export model (the platform would eventually be supplied to the Indian armed forces) are counted as offset, but not as deemed exports. For example, if an aircraft manufacturer OEM is getting the aircraft assembly done at HAL for final supply to IAF, then the supplies made by private Indian defence industry for the assembly process, even if paid in foreign exchange by the OEM, will not count towards the offset obligation fulfilment as goods have not left the domestic territory of India. Such a condition is restrictive and acts as a disincentive for the OEM to develop the local vendor base.

It is important that the MoD recognizes the offsets business as 'deemed exports', since the transaction is in FE and the product is sourced by the foreign OEM. It is only that the products so supplied are delivered within the domestic territory. This change will greatly help in many companies discharge offsets in greater quality and quantum, as well as encourage participation of lead integrators in many programmes, which in fact contributes to a greater value addition in India.

FDI IN DEFENCE MANUFACTURE

The government of India had earlier allowed for 100 per cent participation of the private sector in defence production with FDI to the extent of 26 per cent subject to licensing from the DIPP.

The FDI ceiling has now been enhanced by the government to 49 per cent. The rationale provided is that such a move will go a long way towards enhancing the technical capability of the private sector in the shortest possible time frame, as foreign OEMs will find it more rewarding in sharing technology. It is noteworthy to mention that cumulative amount of FDI in defence sector as of 2009 stands at a mere $4.16 million.

There is also a significant requirement for clarification on what constitutes defence-related industrial infrastructure. This is especially needed with regards to land and related assets that a defence-oriented business may acquire in its efforts to establish a manufacturing base and this aspect needs further clarification.

FDI is the key to a business model and with an enhanced FDI, the foreign OEM has a sustained interest in the Indian partner and is likely to nurture the production to mutual benefit. As long as 'mutual benefit' is the central theme, offsets will grow. In offsets business, the important stakeholders are the OEM and the

IOP and both need to realize the mutual benefits for the business to be sustained.

FDI IN DEFENCE MANUFACTURING AND R&D

Investment in defence R&D is currently allowed but such investment has to be routed to an agency approved by DOFA. Since offsets are purely a commercial transaction and that the OEM has the moral and financial obligation to meet, it is only fair that the process of identification of a suitable partner destination for FDI in defence R&D be left to the foreign vendor.

In line with the above, there is also the requirement of establishment of a separate fund to provide necessary resource to public/private sector, academic and scientific institutions to support R&D for upgrading of different platforms/systems and to promote innovation.

Till date, despite more than US$5 billion signed in offsets contracts alone, there is no single case for FDI in R&D. It is thus obvious that there is something wrong with the formulation of the policy, which has to be addressed by the MoD. This has not been adequately addressed in the policy for offsets 2012 either. The government, by placing 'Services' under abeyance, has done a disservice to the services sector. One default cannot be enough reason for the government to take such a harsh decision on the services sector, which is in any case the backbone of the industry. This needs immediate review, and such knee-jerk reactions of adverse nature could be avoided.

TECHNOLOGY TRANSFERS AND OFFSETS

Technology transfer significantly impacts the growth of Indian defence industry as this is the most significant driver. The Indian industry needs technology for its global acceptability. It is the lack of technology with the Indian industry that is driving the armed forces to look beyond our frontiers for procurement of weapon systems. The practice of nominations (only OFB and DPSUs) by the Department of Defence Production (DDP) has not borne much fruit in the past 60 years or so. Maybe it is time to look beyond this practice of nominations and truly create an atmosphere for a level playing field. The case in point is the IAF's programme of the medium transport aircraft, which reportedly is a deviation from the existing practice of nominations. In this case, the programme is categorized as 'Buy and Make with ToT' and the request for proposal (RFP) is issued only to the foreign vendors of repute, while they are free to choose their Indian manufacturing partner.

Once the nomination of production agency is done away with, technology transfers could take place with any company in the Indian defence industry, paving the way for an accelerated growth. In such an atmosphere, true technological transfers could take place.

BANKING OF OFFSET CREDITS

DPP 2006 provided that only contracts for export of defence products or services or investment made after the signing of the main contract would be reckoned for discharging offset obligations. This was a limitation as it did not provide enough flexibility to either the supplier or the obligor on whom the offset obligation resided in planning offset oriented off-take. In DPP 2008, these constraints were somewhat relaxed and provisions allowing foreign vendors to create offset programmes in anticipation of future obligations and banking of offset credits from such programmes were introduced. Prior to current revision, a vendor was able to discharge the banked offset credits for the RFPs, which are issued within two financial years from the date of approval of the banked offset credits.

Having a period of validity of two financial years proved to be a disincentive for foreign vendors who have singular exposures or where the next programme exposure for a foreign vendor is distant. In such cases, the foreign vendor is forced to defer offset transactions and push them as close to award of next contract as possible thereby diminishing the impact of the banking of credits regime. Therefore, the government increased the validity of the banked credits thereby incentivizing the foreign vendor to commence offset business in India despite having little or no visible programme exposure in the near future. *This has since been revised to seven years, which may be still inadequate.*

Banking of offset credits will now be handled by the newly created DOMW, the Defence Offsets Management Wing.

Procedures laid down for processing of banking proposals need to be implemented more vigorously, as are being done with the procurement proposals. These could be well linked to the Defence Acquisition Council (DAC) and the DAC be apprised of the progress of proposals received in the preceding quarter.

Proposals for banking of offset credits could be more liberal than the procurement cases.

The new policy has also provided a window to the OEMs to propose a banking proposal for all the investments they have made since September 2008 by March 2013. DOMW will be the single window agency to handle all of these.

TREATMENT OF HOMELAND SECURITY AND CIVIL AEROSPACE

For small- and medium-sized manufacturers, defence manufacturing is characterized by large investments, moderately planned production off-take and long development periods. Additionally there is significant commonality between needs of some segments of homeland security and civil aerospace manufacturing with defence manufacturing. For example, an aircraft engine manufacturer will have largely the same product line for defence and civil applications. In such an industry, where placing orders with defence industry provide offset obligation fulfilment to the OEM; while civil applications and products of dual use, do not yield any offset benefit, the economic incentive that can be realized for the IOP does not always accrue. This is again an issue that the government may want to look at to make the offset policy more robust with a larger pool of benefits for the industry at large. In this context, addition of civil aerospace and inland security in the ambit of eligible products is a very welcome step. This could also be expanded to include strategic sectors, such as nuclear and space.

DEFENCE SPECIAL ECONOMIC ZONES AND OFFSETS

There is a requirement for the MoD to aggressively step in and provide for Defence Special Economic Zones (SEZs). There are a number of private industrialists who could team up and create world-class manufacturing capabilities in India. This needs to be encouraged in a more systemic manner. Under the cover of the secrecy/holy cow status or the fear of repercussions from the unknown (since MoD has not been able to come out with an explanation why they are actually against this proposal), the MoD is probably not venturing into this area. While the Ministry of Commerce is aggressive in promoting SEZs, the MoD is seen to be taking a retrograde step. In order that we achieve our objective of self-sustenance, which is intimately dependent upon aggressive growth, hard core manufacturing, nurturing talent in design and development process, etc., the government must take the lead in promoting such manufacturing parks that may create an ecosystem for strategic manufacturing. These are such small initiatives that will pave the way for growth. It is a very good idea to have dedicated manufacturing zones established for synergy and growth and the existing SEZs concept can be exploited to its optimum. There is a positive FE flow when the SEZ entity supplied to the government agencies such as armed forces, DRDO and paramilitary forces. Also, this would stimulate exports, ethos of SEZ remaining a priority for the units located there.

Locating a manufacturing or services unit in an SEZ also comes with the twin advantage of having a single window for licensing under the Board of Approval (BOA), which is governed by the SEZ

Act, besides the advantages of taxes and a whole lot of incentives provided by the State and the Centre.

DEEMED EXPORTS

All transactions from offsets must be notified as 'Deemed Exports' and maximum possible export benefits be provided to them. The offset transactions do invite FE (positive FE), while the port of delivery is within India in many cases. Thus real export does not take place, although the foreign OEM has paid for it in FE. The Indian industry especially those located within SEZs will immensely benefit if such status be accorded for all transactions necessitated by offsets.

SEZ AIRPORT

There is a requirement of an SEZ airport. This will cater for the huge number of aviation orders that are in the pipeline. The foreign OEM could place his aircraft for integration in an SEZ airport and the Indian industry could avail of all the benefits of exports for all the integration work or supply of components and sub-systems for the target aircraft. The government must create an atmosphere for creation of SEZ airports.

USE OF MULTIPLIERS

The offset policy of India did not allow the use of multipliers, which are a device to give additional credits for discharge of offsets in a directed manner—either through industry segments or through critical technologies, till recently. However at this juncture it is important to note what has happened elsewhere in the world with regards to multipliers. A case in point is Turkey where foreign vendors can earn multipliers for specific technologies, first time exports and even working with small and medium enterprises.

The use of multipliers will further help in directing the development of the indigenous defence industry to pre-determined segments. This appears to be less of a danger than allowing the development of defence industry, whereby, India may remain content with having the capability to produce small components of defence equipment, but not the complex systems. Considering the pace at which Indian offset policy has evolved, it may be prudent for the MoD to explore the benefits that multipliers can bring to Indian defence manufacturing industry. Multipliers are required only when the priorities are clear (areas where offsets are preferred) and the requirement is overpowering. This situation is less likely to arise in our context and multipliers may not find the necessary grounds to find a place in the policy. In recognition of the advantages and possible benefits that multipliers can offer, it was felt necessary to incentivize the OEMs for any technology that they may transfer. Accordingly, the MoD in the recently released revised offset guidelines, have catered to a multiplier of up to 3, in the context of technology acquisition, and a simple multiplier of 1.1 for a buy-back, in case of manufacturing ToT.

Multipliers have also been accorded to OEMs who pick up MSME units as IOPs. This will encourage the OEMs to have a closer look at the MSME segment in the country.

MSMEs AND OFFSETS

Defence sector is a strategic sector and has been a protected sector for a long time. The first movement took place in 1991 when the National Industrial Policy was announced, which made a major transformation to many sectors including defence. Defence sector was removed from the 'reserved category' and placed in the 'licensed category'. This truly was the beginning of inviting private participation in the defence sector. Prior to this the sector was a holy cow, under the public sector alone, dominated by the OFB and the DPSUs.

Despite such a radical change in the policy announcement, there was not much of a movement in terms of either participation or involvement of the private industry. The major movement actually took place after the government revised the sector-based caps placed on the different sectors through a press note in 2001, whereby the defence sector was opened up to 100 per cent participation by the private industry, with FDI investment up to 26 per cent, both subject to licensing. Detailed guidelines were promulgated by the Ministry of Commerce in 2002, again vide a press note for participation of the private industry in the defence sector. This marked a major movement from the side of private entrepreneurs for participation in the defence sector.

While all this was happening in the Ministry of Commerce and Industry, the MoD in parallel was putting in place a mechanism for procurements in defence, which is hitherto popularly referred to as the DPP.

PROCUREMENT SYSTEMS AND PROCEDURES IN INDIA

Before we attempt to see in detail the procurement procedures at the ministry levels, we must have a look at the General Financial Regulations 2005, popularly known as GFR 2005, that lay down the general principles for every government procurement agency to follow. The MoD followed it through till they put DPP in place.

While the GFR is a general guideline, the concerned procurement agencies of the government, within each of the ministries, may lay down their own procedures so long as they are in conformance with the GFR 2005. The section on procurement of goods lays down the fundamental principles of pubic buying. The stages indicated are such as laying down of the qualitative requirements (QRs) in a concise manner without being superfluous and plan procurement of only the quantities that are required. This must be done by an invitation of offers in a fair and transparent manner with adherence to reasonable procedure. While the procuring authority must be satisfied of the general adherence to the requirement specification while making the procurement, they must also establish the reasonability of price with consistent quality. A single-stage, two-bid system with the technical and commercial bids in separate envelopes is suggested. Maintenance contract has been encouraged for large systems to ensure operational status and an advance payment of up to 30 per cent of the contract value has been allowed.

The DPP is an elaborate document with details of the extensive process followed by the MoD. The DPP is placed in conformance with the GFR 2005 and has indicated the details of the methodology in planning and execution. The DPP also indicates a single-stage two-bid system as advocated in the GFR 2005.

The first document was that of the DPP 2002 and since then it has been evolving year on year based on the experiences of the MoD in dealing with various foreign vendors. DPP 2011 spells out the various models followed by the MoD for procurement of weapon systems, in capital procurement. The other document of relevance in our context of the MSMEs is the Defence Procurement Manual (DPM), which relates to the procurements made from revenue stream.

MSMEs till recently did not find a place in the DPP. The first sign of inclusion was in the form of a multiplier accorded to the OEMs for choosing an MSME as IOP in their offset proposal. There is a need for the DPP to address the MSME segment more poignantly and make a deliberate inclusion in the main procurement procedures itself, both for capital and revenue acquisitions (DPP and DPM).

MSMEs IN STRATEGIC ELECTRONICS SECTOR

The strategic electronics sector is well placed vis-à-vis the other sectors. The applications are very diversified and find applicability in almost every weapon system. It is estimated that electronics have a share from 36 to 66 per cent in every major weapon system both in aerospace and defence. Therefore, the electronics industries with niche capabilities are already diversified in application and need to harness the opportunity available.

The electronics industry, however, needs to follow the indications as available in the request for information (RFI) process that is now being followed by the DPP. Even before any major acquisition is to be planned the MoD issues a RFI to be placed on the MoD website or any of the government procurement websites providing an indication of the procurement. At this stage, the industry also gets a broad indication of the procurement to follow, and there is adequate lead time available for the industry to prepare for the ibid procurement. Thereafter usually there are a number of seminars and conferences organized by the industry associations where most of these acquisitions are discussed with the stakeholders in attendance.

PLANNING PROCESS

The rest of the planning process involves the formulation of the QRs that actually flow from the indicated operational requirement of the lead service. The industry once again has an opportunity to get involved in the formulation of the QR (called as General Service Qualitative Requirement in the case of Army, NSR for Naval procurements and ASR for Air Force), through the medium of the industry associations who are called by the MoD for their comments with respect to any of the parameters in the formulation of the QR. In fact, the response to the RFI is an important indication for the armed forces to estimate the level of seriousness of the industry in participation in the ibid tender. A serious industry player can impact the formulation of the QR in a more positive manner.

The planning process thereafter is confined to the preparation of the briefings, discussions in the various capital acquisition committee meetings, the lower and the higher committee before it gets a final nod from the DAC in terms of the Acceptance of Necessity. In this process, once again the industry is invited by the MoD and the HQ IDS, for internal briefings and discussions with regard to the capabilities existing in the Indian industry. This is a good chance for the industry to showcase their capabilities to the MoD. This will have a huge impact on the decision as regards to the categorization of the ibid system, either to go the 'Buy Indian' way or the 'Buy Global' way. In the case of 'Buy and Make Indian' category the Indian industry lead must be able to contribute at least 50 per cent of the value by cost in the aid acquisition, while they are free to have global tie-ups with the OEM of their choice. Thus, the MoD has very deftly done away with the nominations in terms of the production agency, which was a necessity in the case of 'Buy and Make with ToT and licensed production'. This level playing field provided by the MoD must be exploited by the electronics industry for their benefit.

EXECUTION PROCESS

The execution process follows the planning process with the formulation of the RFP; the consequent vetting in two stages and shortlisting of vendors to whom the RFP must be issued. This is again a critical stage for the industry; since, once the RFP is issued there is no way that any new vendors can be added, even if there is an omission. So what must an industry do to guarantee that they get the RFP?

The industry has to be involved from the response to the RFI, subsequent discussions with the MoD and continuous engagement, participation in the formulation of the QR, discussions with the planning departments of the various services, engage with the user (the procuring directorate), discussions and engagement with the HQ IDS through the medium of the industry associations and more importantly registration.

REGISTRATIONS

There are typically two types of registration spelt out by the DPP, one is at the service HQ level and the other is with the DIPP. The registration with the services needs to be done with the respective Technical Manager (TM) concerned and the other is with the Ministry of Commerce, DIPP, and Secretariat of Industrial Assistance (SIA). The one with the TM in the respective service is fairly easy which is a one page form giving out the details of the industry, product and basic information of the company. This can be done online as well as directly by submitting this form with the office of the TM. The TM, directly functions under the Director General Acquisitions, and is actually responsible for the issue of the RFP. Therefore this registration is important for a company to receive the RFP. It may also be important to be visible to the ultimate user, in this case the line directorate or the procuring directorate, since they are the ones who do the first stage of vendor selections. There is no formal registration process set up here; however, it is useful to be in their notice. This happens from the response to the RFI stage and thereafter a continuous process.

EVALUATIONS AND TRIALS

Once the RFP is issued and responses are received, in accordance with the single-stage two-bid system, the entire process of evaluations, trials and shortlisting is followed. If the system under procurement is of a restricted nature, then, the MoD will first confirm if the vendor is holding a valid industrial licence for the same. This is where the DIPP comes in. DIPP is the single window for licensing across all industries in India. The concerned entity must submit the details as asked for in the forms that are freely available on the ministry's website (www.dipp.nic.in), and submit the required number (16) of copies of the application form with a token fee (₹2,500) by a demand draft to the SIA, DIPP, which has its office in Udyog Bhawan. The DIPP follows up with the respective user ministry, in this case with the MoD, by sending the requisite details for further examination. The MoD conducts regular meetings of the licensing committee, which is a multi-disciplinary committee, comprising members from the services, the HQIDS, DRDO, DGQA, MoD and then they take a decision if the IL can be given to the said entity. This is communicated back to the DIPP, who then follow the recommendation of the MoD and issue the said IL.

OFFSETS

In addition to all of this, there is also an opportunity in offsets. The opportunity is real and large; hence, there is a business waiting in every discipline in the entire industry segment. This means that a PCB manufacturer is as much at an advantageous position as is a sub-system manufacturer or those who make large systems like radars and integrated systems. Every small-and-medium-sized enterprise in this segment has a very important role to play, with a little help in terms of certifications, best practices, necessary quality assurance checks and financial support.

So, what does an industry/enterprise do, if they are looking for a participation in offset programmes? If in case of the above licensing process, the said item does not fall under the conditions for licensing, then the DIPP issues a letter to this effect. This letter is also important, to let the OEMs know that such a requirement does not exist, so that the said enterprise is a valid IOP. The conditions of licensing for IOP have been done away with since the DPP 2008 (which was promulgated as a compulsory condition in DPP 2006) and the conditions are more relaxed and favourable.

LICENSING CONDITIONS OF DIPP

So, what are the licensing conditions of the DIPP? The industrial policy of 1991 is very generic in its conditions such that electronic equipment of all types fall under compulsory licensing. When the defence sector was opened up in 1991, the industry was actually removed from the reserved category and placed in the licensed category, subject to licensing. The government also removed restrictions on the FDI, by opening up the industry to investments and allowed a maximum of 26 per cent FDI, both subject to licensing (both the quantum of investment and the enterprise). So if an enterprise is looking to be an IOP to a foreign OEM to participate in the offset contracts, then the said enterprise must follow the twin conditions, as laid down in the press note 2 of 2002. These are not having an FDI of more that 26 per cent in the enterprise, the management be completely Indian, including the CEO, and the product be licensed (should fall under the restrictive regime).

RESTRICTIVE REGIME

So, what is this restrictive regime and how does one come to know if the product to be manufactured is indeed a restrictive item? For this, one must refer to the International Trade Classification, harmonized Code (popularly known as ITC-HS). This gives a very clear indication of the nature of the product with the cat part number and the restrictions associated with it. A further reference is the SCOMET list under the export category.

REGULATORY ASPECTS

The MoD is interested in procuring the weapons and systems that the armed forces require in a manner described in the DPP. The regulatory aspects come into play when a short-selection is to be made for the issue of RFP vendor validity and reliability or for the choice of the IOP. The OEM is actually responsible to make sure that the IOP selected by them are in conformance with the rules and regulatory aspects as specified in the various instruments of legislation in India. Besides all the aspects of licensing and conditions that govern this choice, the actual aspects that are important are the capabilities of Indian enterprise for manufacturing such a product and the various certifications associated with it. The Indian enterprise must necessarily have all the certifications governing the manufacture of the said product. Especially, as it concerns the OEM, adherence to the global best practices will be important for quality adherence and capacity production in the time frame required.

SERVICES INDUSTRY AND OFFSETS: RULES AND REGULATIONS AFFECTING THE INDUSTRY

Services industry have been freed from the system of licensing and even going by the extant regulations of the Government of India, up to 100 per cent FDI is allowed in the industry. The industry is also not governed by licensing requirements and IL is not applicable to them. However, the MoD is following the spirit of the DPP to develop the defence industrial base in the country through the medium of offsets. The industry has been designated as an Indian industry in conformance with the DIPP guidelines as laid down for the manufacturing sector. If the service industry wants to participate in the offsets opportunity as an IOP, then it must be in conformance with the guidelines as laid down for the manufacturing sector as stated and described earlier.

An analysis of the offsets discharge in the recent past is indicative of the dominant role played by the services industry. This includes, for example, MRO, simulators and training, testing and evaluations, IT and design services and military-related software.

WHAT ARE THE REGULATIONS FOR PARTICIPATION OF INDIAN INDUSTRY DIRECTLY WITH THE MoD?

To be eligible to receive an RFP from the MoD directly either as part of the revenue stream or as part of the capital stream, the governing conditions are as given in the DIPP. This means that a wholly owned subsidiary of the foreign OEM duly registered in India is eligible for participation in all contracts. In this case, the cap of 26 per cent on FDI is not applicable. However, the licensing conditions as applied to products falling under restrictive category are applicable here.

The policy and regulatory conditions are very conducive for the participation of the Indian private industry as is for the public sector. The erstwhile nomination-based single-source procurement is no longer being practised by the MoD (this may be adopted in very exceptional cases, with due justification). Therefore, the opportunity lies in participation into the various MoD contracts directly in the 'Buy Indian' category, or with foreign linkages in the 'Buy and Make Indian' category. In the case of the 'Buy Global' and the 'Buy and Make with ToT followed by licensed production' category, the participation is in terms of being an IOP with the selected OEM. In the 'make' category, it may be prudent to have strong linkages with the lead system integrators and be a part of their supply chain for the various opportunities there.

INDIGENIZATION

The Indian army holds a sizable number of equipment of foreign origin. During the initial procurement, only limited quantity of spares are procured along with the equipment. However, by the time the equipment is exploited in the field, spare support from the OEM dwindles, as production lines are generally closed or are diverted to other products due to rapid changes in the field of technology. The aim of the indigenization programme of the Indian army is to develop major systems and sub-systems such as propulsion plants, prime movers for power generation, hydraulic systems, auxiliary system, electrical and electronic systems for the vehicles, weapon system and other military systems. The major steps involved in the indigenization process are identification of items for indigenization, benchmarking of price, generation of drawings/QA specs, identification of vendors, RFP, technical evaluation of bids, conclusion of contract and post-contract management.

DPM 2009 has major enabling provisions for encouragement of the private sector. Development contracts will be concluded as far as feasible with two or more contractors in parallel, subject to other vendors matching the price of L1, ratio of splitting of order between various development agencies in cases of parallel development will be disclosed in the RFP itself. To ensure an economically viable quantity, minimum of five years of requirement will be taken while placing the development order. There is a provision for placing orders on single source also, subject to the technology being complex and expertise not generally being available.

NSIC: CHAMPION OF THE MSMEs IN INDIA— ROLE OF NSIC

NSIC is a Mini Ratna PSU under the Ministry of MSME, Government of India, established in 1955 to champion the cause of the MSMEs in India, in all disciplines. Recently, NSIC has entered into the defence, internal security and civil aerospace sectors, with a specific focus to bring the MSMEs into this business and showcase their capabilities to the stakeholders, such as the MoD, DRDO, Army, Navy, Air Force, Coast Guard, DPSUs and foreign OEMs for offsets.

NSIC by the virtue of the MSME Development Act is considered as a manufacturer. NSIC can receive RFPs, bid for them and in turn help MSMEs get good business. The entire BD for this is undertaken by NSIC and for its registered members; all provisions as applicable to the MSMEs will be granted. Adequate information will be made available to the enterprises and they will be encouraged to take part in the process. NSIC can also form consortiums for development of certain specific sub-systems and take part in the entire acquisition process.

NSIC can do the following for MSMEs:

(a) Enable MSMEs to participate in the 'Buy Indian' process of the capital acquisition process.
(b) Enable MSMEs to participate in development programmes of MoD.
(c) Enable MSMEs to participate in the indigenization programmes of MoD.
(d) Enable MSMEs to participate in the offsets programmes by pairing up directly with OEMs.
(e) Find opportunities in synergistic sectors for MSME participation.

KEY ISSUES: IMPLEMENTATION OF OFFSET POLICY

Offset policy is a bold step in the right direction by the Indian government. However, this opportunity catalyst is significantly linked to the easing of the overall procedural constraints. For the full impact of this policy to be felt, it is important to address key issues. Given below are some of the key issues faced by the industry in the implementation of this policy.

DEFENCE OFFSET MONITORING WING

The DOFA was created as a central node to address issues related to offset development. While DOFA has performed creditably since its inception, it is only now that the effect of the policy is started to being seen. It is imperative that the government equip the newly announced DOMW with adequate capability to not only address the needs of the industry today but also for the future. The establishment of a single point of accountability designated as an authority, for the entire offset process is extremely critical. This is needed to ensure that one nodal agency within the MoD be responsible for the evaluation of offset proposals as part of the acquisition process, approving projects during implementation and approving and banking the offset credits gained for that project.

Having an offset organization in place that is structured to support the acquisition process and to manage the successful implementation of the resultant obligations with the OEMs can go a long way in ensuring a successful offset regime. The proof is, however, in its effective implementation.

EVOLUTION OF SOURCES OF STRUCTURED FINANCING FOR THE INDUSTRY

The growth of Indian manufacturing capability as a response to the offset opportunity is also a function of the ease of availability of finance. Traditionally, the defence industry is characterized as being cyclical with long gestation periods, high development costs and low repeat order possibilities. Hence businesses, especially in the MSME segment in this industry, typically face a credit crunch as these businesses have low attractiveness and high-risk exposures. The development of sources of structured financing in this industry will go a long way in sustaining the Indian industry's capability to make full use of the offset opportunity.

TRAINED MANPOWER

A large business opportunity as provided by offsets requires shoring up of pools of talent to be able to create and run businesses profitably. India has only just started witnessing an interaction between the pool of manufacturing talent (which primarily resides in the DPSUs, DRDO and private industry) and the users (from the armed forces). This emerging mix is providing the right human capital framework to create businesses that can take this high growth phase forward.

EXPERTISE IN MANUFACTURING

Manufacturing cannot be a subset of offsets. In fact, offsets should happen because there is manufacturing. The process involves investment, certification, quality control and host of other issues. Offsets can be utilized over short term to widen the manufacturing base, but core competency in manufacturing will come out of the policy of 'Buy Indian' and 'Buy and Make Indian'.

ORIGINAL EQUIPMENT MANUFACTURERS/SYSTEM INTEGRATORS

Globally, the manufacturing ecosystem consists of the following tiers:

(a) **Tier 1**

    i. These companies manufacture major sub-systems and structures of the overall system.

    ii. They too do not manufacture each of the components required for their respective systems.

    iii. The larger and well-established Tier-1s are generally seen to be a part of the OEMs or were so at some point of time, though this is not absolutely necessary.

    iv. Examples of such companies are Spirit Aero systems; Martin Baker; Hamilton Sundstrand; QinetiQ; Pratt & Whitney.

(b) **Tier 2**

    i. These companies manufacture sub-components/sub-assemblies for the Tier-1 companies.

    ii. They execute work packages such as Precision Machining; Sheet Metal; Casting; Forging; etc.

    iii. Examples of such companies are Meggitt Systems; Marshall Aerospace; Magellan Aerospace; Premier Precision.

(c) **Tier 3.**   These companies carry out specific processes such as Welding, Surface Treatment, Non-Destructive Testing and Engineering Design.

In India, the distinction between tiers is pretty fuzzy. Besides the lead integrators, which are the DPSUs and the top companies, there are only SMEs. India should develop a tiered structure, with some companies coming in between the OEMs and the SMEs. Such companies need to be encouraged by the government by means of suitable incentives

GROUND REALITIES IN VALUE CHAIN IN INDIA

In the defence production life cycle, India is mostly getting 'built to print'. The need of the hour is to graduate towards built to specifications. Currently, the entire value chain is controlled by the OEM who decides the various stages of procurement. The issues that should cause concern for Indian manufacturing are:

(a) Can we manufacture military-grade steel in this country?
(b) Can we manufacture titanium products in this country?
(c) Are we looking at establishment of a titanium plant?
(d) Can we make a titanium sponge?

If we do not do all this and only do contract manufacturing, the country cannot graduate to the next level. As on today, we have limited competency, with capacity in machining defence electronics, small gearboxes and similar products. We lack competency in bigger aspects like raw materials, castings, wind tunnels amongst others. We need to build success stories of Indian companies getting into the global supply chain and develop further on these. That will sustain the offset initiative towards a long-term manufacturing industry in this country. From an analysis of figures that have been available in the public domain, a major portion of the offsets is going towards services. If offsets have to lead the country to improving manufacturing, their share in manufacturing has to be enhanced. There are indeed challenges like the requirements of ILs, higher attractiveness of the service sector, lack of competency in high-end manufacturing amongst others. Heavy investment is needed in infrastructure; the policy really does not support this. We talk of value addition and manufacturing ToT, but we do not incentivize the OEMs to provide manufacturing ToT. There are questions of certifications, penalties and the risks and of course the other things going with it. Let us take the case of an OEM who wishes to select an IOP. There are three possibilities.

(a) Possibility No. 1: Manufacturing capability exists in the country.
(b) Possibility No. 2: Manufacturing capability exists partially.
(c) Possibility No. 3: Manufacturing capability does not exist.

So if an OEM wants to engage an Indian partner in manufacturing, obviously he will take Possibility No. 1 where the capability exists. In India, this capability exists only in low-end technology. Hence the OEM is handicapped in case he is required to provide high-end technology. If the OEM wishes to put up an industrial infrastructure for the IOP, there is no incentive/support from the government. The myth of India being a low-cost manufacturing hub does not hold good too due to issues of economy of scale. Then there is a question of due diligence, investment in the infrastructure, requirement of a large number of so many certifications that you need to have, etc. The policy provides the freedom of choice to the OEM to select the IOP basically because the government does not want to bear the risk, should the IOP fail with the OEM having to pay a penalty for delays in delivery. The OEM has to make changes to his supply chain. Now, why should he make changes to the supply chain and choose an Indian partner and bring him into the supply chain and enable him and give him certifications, get him the quality assurance, do the testing and evaluation so that the manufacturing in India benefits both of us. Where is the incentive for him to do all this? The policy in principle is not supporting, enabling of manufacturing technology to come up. However, with the visible multipliers accorded in recent modifications to policy, may be there could be some forward movement.

CONCLUSION

The DPP today includes 'Buy Indian', 'Buy and Make Indian' and 'Make' categories for procurement. The future is going to be in the direction of 'Buy Indian', 'Buy and Make Indian' and 'Make'. There is going to be a gradual reduction in 'Buy Global' and 'Buy and Make with ToT'.

If that were to be the case, R&D is going to play a very important role. Yet we do not have a system in the Defence Procurement Process for issuing an RFP for design and development. The country needs to focus on R&D in the defence as well as private sector if it is expected that offsets would make a difference to our manufacturing capability.

If the country lives in the villages, the industry lives in the MSMEs. NSIC is already championing the cause of MSMEs in India and they need to play a more active role in the defence planning and implementation process for a more aggressive role for the MSMEs.

9

Knowledge Arbitrage through Defence Offsets

S.S. Mehta

Abstract

The author debunks myths related to defence offsets. He is of the opinion that defence offsets if employed gainfully can help India change its profile and upgrade from labour arbitrage to knowledge arbitrage.

INTRODUCTION

India is one of the most lucrative markets for military products in the world. India's industrial policy has by and large kept defence production sector under government control. The tools for bolstering indigenous defence production capability were transferred through licence agreements and technology transfer to Defence Public Sector Undertakings and Ordnance Factories. While these methods enhanced the levels of technology, the fact is that even today over 70 per cent of our defence acquisitions are from external sources.

The Ministry of Defence (MoD) has only recently but slowly and surely moved towards an offsets regime, which has galvanized public, private and foreign suppliers. The outcome is difficult to predict. It must, however, be said that this is a welcome initiative.

Much learning is required for its effective execution. It is a global practice and there are some very good and some bad experiences from which we need to learn.

The initiative is timely because we are in the midst of a unique demographic advantage that provides us not only an opportunity but also a challenge. While the opportunity lies in our abilities to harness the energies of the youth, the challenge is to create job opportunities and skills in our work force so that they can benefit from greater and enhanced livelihood prospects.

It is axiomatic that an engaged and productive workforce, while satisfying aspirational needs of individuals and families, should also provide competitive advantages to our industry, at a time when the developed world and China are ageing and growth levels in developed countries are crawling at abysmally low levels as never before. The other and equally important advantage is that it ensures higher levels of national security. It need not be emphasized that jobless youth are crucible for mischief. We have already had enough of that.

In the achievement of this objective, there is a case for a *National Mission* for moving India from a 'labour arbitrage' economy to a 'knowledge arbitrage' economy. Defence offsets are one of the tools for the fulfilment of this objective. The favourable effects of strategic defence spending will show up in higher levels of security, training of manpower, improvement in border and manufacturing infrastructure, in enhancing levels of 'job quantity' and, more importantly, in 'job quality'. Offsets should help create a vibrant and a strong industrial base and accelerate our indigenous capability in producing world-class sub-systems and platforms. It is our national interest to move away from 'net importers' of defence equipment to 'reduced levels' of imports, and, opportunity permitting, to export to those democratic countries that have a secular and multicultural polity.

Brauer and Dunne (2006) in their paper in Economics of Arms Trade Offsets note that

> Offsets, arrangements that obligate the arms seller to reinvest ('Offset') arms sales proceeds in the purchasing country, are an increasingly important

part of the international trade in arms. When countries procure defence equipment from a foreign supplier they look to reduce the costs in a number of ways. They may become involved in the development and production of the product, e.g. joint production, licensed production, or sub-contractor production. Foreign direct investment, technology transfer, and counter trade are other methods of compensation, which may take place in the civilian rather than military sector. Each form of involvement—lumped together under the concept of 'offsets'—carries its own implications for costs, programmers, control over specifications and wider industrial and economic benefits. (Hartley, 1995)

Countries apply different criteria for determining whether offset obligations are required for a particular transaction and what types of offsets are acceptable. While a generally agreed-upon definition of 'offsets' has yet to emerge, by way of illustration, the US government defines offsets as 'Industrial compensation practices required as a condition of purchase in either government-to-government or commercial sales of defence articles and/or defence services'. In principle, offsets hold out great promise for developing and emerging economies. They can spend their budgets on arms and along with it, upgrade their industries, both through developing their arms-related industries and other related and unrelated/ ancillary industries.

It is necessary to map the global experience of offset objectives set out by the arms importing states. Brauer and Dunne's analysis in this regard is instructive. They opine that

Some states target certain arms niches that they wish to learn to master for themselves and they structure arms import acquisition and offset demands toward the fulfilment of that goal (e.g. Singapore, Taiwan). Other countries have well-developed specialized arms production niches and use arms trade offsets to assist them to maintain international competitiveness in that niche. Their objective has evolved from vaguely promoting general economic development to the development and maintenance of specific arms production competences (e.g. Sweden, the Netherlands). Still other countries (e.g. Brazil, India and Indonesia) appear to be driven by regional power ambitions that would dictate the development of an indigenous ability to produce sweeping plate of weapon systems in country and they therefore pursue an arms sourcing and offset strategy with broad transfer requirements. Yet other states (e.g. South Korea) seek an ability to produce a wide spectrum of systems not because of regional power ambitions but because of a generalized desire and increasing ability to broadly participate in all industrial markets. Still other states appear to view arms offsets as an opportunity to revive a collapsed or failed indigenous arms industry (e.g. Poland). Other states (e.g. the U.K.) view offsets as a tool towards reaching an ever more advanced state of a globally integrated arms manufacturing system in which producers residing in various states produce components for sometimes this, sometimes that lead-manufacturer. (2006)

In contrast, India has set out very modest and non-threatening guidelines in its recently released offset policy. The key objectives of the Indian Defence Offset Policy is to leverage capital acquisitions; develop Indian defence industry by fostering development of internationally competitive enterprises; augment capacity for research, design and development related to defence products and services and to encourage development of synergistic sectors like civil aerospace and internal security. All these are laudable objectives and signify a clear message that India is on the learning curve, and will, over time, calibrate and fine-tune what is best for India. These objectives will, on one hand, help to achieve higher levels of self-reliance in the critical defence sector, and, on the other, create the much needed manufacturing infrastructure in synergy with Public Sector enterprises—all adding up to better skills, more jobs and provision of a boost to preparing our industry against global competitors.

Some prevailing myths need to be addressed. Colin Greer in his article titled 'The Biggest Engine of Growth' addresses two perennial myths.

Those who claim the superiority of private capital and insist that government is not as an inventor or venture capitalist should consider the history of the jet engine, the computer and the internet to name just three inventions that have been essential to progress and technology growth. They were all developed in a government laboratory in the late 60s, and their early applications were heavily underwritten by the government. (2012)

Another myth of the market capitalism narrative that Colin Greer debunks is that 'innovation comes out of market competition'. This notion, too, is tossed on its head. As Shellenberger and

Nordhaus write: 'Today's relatively inexpensive jet travel began with Pentagon procurement and research and development (R&D) for jet turbines in the 1940s and 1950s. But it took many decades before jet travel became accessible to the average American and much of the rest of the world'. This is perhaps the most unseen of the government's functions. As a silent partner, the government brought massive capital investment to the advancement of technological research and development with no return on investment beyond tax revenue growth to capture it. Phones, radio, TV, computers, the satellite system, nylon (invented in place of silk as war with Japan loomed), Velcro and breakthrough drugs were all the result of intense public investment in university and corporate research and development.

R&D is thus a key aspect of the offsets package. It must remain and be monitored as impeccably as the other components of the offset package. Since these have long gestation periods and are invisible in the public eye, they tend to suffer from the consequences of poor patronage. More so in an environment where a regular budgetary squeeze for the genuine demands of the social sector is a 'given'.

Gregory Martin in his paper 'Offsets: Sharing International Experience' makes some candid observations. He says,

> When effectively managed, offset is an opportunity for both. For the Buying Country, implementation of an offset policy can introduce investment, R&D, new process techniques, and global market access to its industrial base. For the Supplier Company, a flexible offset plan encourages overseas investments and facilities access to certain in-country resources, including a wider supply base, oftentimes economical manufacturing, a skilled labour pool and new technologies. (2006)

After many halting and, at times, debilitating starts, our MoD needs to be credited with what now appears to be a sound policy. Its value will of course lie in the speed and dispatch with which it is implemented. It is an essential tool in moving Indian manufacturing up the value chain. With it will come all the co-commitment bonuses of a technology upgrade, higher levels of skill, and better and greater employment opportunities. Young India awaits an exponential rise in livelihood opportunities. India now has the opportunity to 'make every penny count towards this objective'. Affordable defence expenditure is one more tool in this value chain.

Strategically conceived and imaginatively deployed defence offsets will change our landscape—from labour to knowledge arbitrage—setting us on the path to higher growth trajectories—all adding up to enhanced levels of national security. It is a win–win equation for all.

BIBLIOGRAPHY

Behera, L.K. 'India's Affordable Defence Spending', *Journal of Defence Studies*, Vol. 2, No. 1, 2008.

Brauer, J. and Dunne, J.P. Seminar Paper: The Economics of Trade Offsets: A Review, International Seminar on Defence Finance and Economics, 13–15 November 2006, New Delhi.

Chowdhry, A. Foreword—Creating a Vibrant Domestic Defence Manufacturing Sector, BCG and CII Paper. Available at: http://www.ciidefence.com/pdf/ Creating%20a%20Vibrant%20Domestic%20Defence%20Manufacturing%20 Sector%20V5.pdf.

Greer, C. 'The Biggest Engine of Economic Growth? 8 Ways Taxpayers and the Government Are Necessary to Capitalism', 13 March 2012. Available at: http://www.alternet.org/story/154538/the_biggest_engine_of_economic_ growth_8_ways_taxpayers_and_the_government_are_necessary_to_capi- talism.

Katoch, R. Seminar Paper: Defence Economics: Some Core Issues and Concerns for India, International Seminar on Defence Finance and Economics, 13–15 November 2006, New Delhi. Available at: http://idsa.in/strategicanalysis/ DefenceEconomicsCoreIssues_rkatoch_0406.

Martin, G.J. Seminar Paper: Offsets: Sharing International Experience, International Seminar on Defence Finance and Economics, 13–15 November 2006, New Delhi.

Martin, S. and Hartley, M. 'Defense Equipment, Exports and Offsets: The UK Experience'. *Defence Analysis*, Vol. 11, No. 1, pp. 21–30, 1995.

Minocha, N. 'Making Things Simpler', G. Files, July 2012. Available at: http:// gfilesindia.blog.com/archives/255/, accessed on 24 November 2014.

Misra, S.N. 'Impact of Offset Policy on India's Military Industrial Capability', *Journal of Defence Studies*. Available at: http://idsa.in/jds/5_3_2011_ ImpactofOffsetPolicyonIndiaMilitaryIndustrialCapability_SNMisra.html.

PART III
VIEWS FROM THE GLOBAL OEMs

10

Offsets: A Global Prime Standpoint

Nalin Jain

Abstract

This chapter surveys the opportunities in defence offsets from the viewpoint of a global prime who has a long-term association with defence supply in India. Offsets can only assist good business plans, not substitute them. Offsets can only be successful if the business plan makes sense. The author talks about the ingredients of a successful offset programme and lays out various models of engagement for the domestic private industry in India. The chapter concludes with suggestions on various options that can motivate global primes to involve themselves in the country in a substantial manner.

INTRODUCTION

In a free market, the whole game of industrialization is about profits. But at the end of the day, what one needs to focus on is that whether the business case makes sense, whether offsets will make sense and not otherwise. The offset programme can be successful only if the environment for business transactions is conducive and all stakeholders can profit from it to make it a win–win situation.

ABOUT GENERAL ELECTRIC

General Electric (GE) is a US$17.6 billion business in 2010 revenue terms. The company is present in all the market segments in aerospace, from commercial engines to defence to business and general aviation, a new market that GE has entered in the last few years and plans for its aggressive growth. Another segment of business is Aviation Systems, which is really avionics, mechanical structures and other integrated systems, which go into an aircraft.

To sum it up, these are all technology-intensive businesses and in these circumstances, GE is the world leader in aviation technology as far as aircraft engines are concerned. In the commercial aviation market, GE has close to 63 per cent of the market and in the military domain, it has close to 46 per cent of the market with significant presence in the fighter aircraft and helicopter space. As far as the aircraft engines are concerned, technologies are very similar for both commercial and military sides.

The volume of business of commercial aviation is high and has become significant for India in recent years. In India, GE is associated with the defence industry at multiple levels. It is with the light combat aircraft programme Tejas since 1986 when the first demonstrative programme was introduced. On the heavy side, the maritime patrol aircraft is powered engines made by CFM in a joint venture with GE. The naval frigates Sayadri, Satpura and Shivalik are powered by LM2500 gas turbines that are GE products. GE is working with the Indian Navy and the Mazgaon dock to integrate these gas turbines into the ships as they get commissioned.

On the system side, avionic packages are present on many of the Indian programmes like Jaguars, R/C Harriers, Hawk trainers and the Seakings. On the mechanical side, the C-130J that arrived recently has GE propellers on them. On the commercial side, GE has close to 450 engines, with close to 60 per cent share of the market. Four of the six airlines are 100 per cent powered by GE CFM engines. GE has also been creating infrastructure in India, for example a training school for CFM engines in Hyderabad at the

GMR airport. A joint MRO facility with Air India in Nagpur is expected to go online in 2013. GE has a 5,000-strong R&D centre in Bangalore, which is one of their largest in the world with 500 people focused specifically on aviation design and technologies.

ENABLERS FOR STAKEHOLDERS

All countries have their own aspirations and goals and reasons why they want to go for offsets. Some do it for jobs, others do it for exports, some others do it for technology and so on. Depending on what their goals are, they can actually tailor their offset policy in that direction. The way we look at our subject of offsets and industrial cooperation means that there are three stakeholders here. The government, which has national objectives of development, self-reliance and so on, private sector in the country, the domestic players who want to develop and grow, and of course the global primes who want to globalize, grow profitably, leverage low cost and become competitive in the marketplace.

The key enabler for the three stakeholders in one form is the environment for their business transactions. The three stakeholders are talking about the environment, effective policy to justify investment and to make sense of the business plan and a smooth entry with minimum hiccups. GE started out in 2005 in offsets in India from the defence standpoint. These are early years, there has been a lot of excitement around the volume of offsets that is going to come into India and ultimately, the consequences it would have for the growth opportunity for the manufacturing industry and industrialization. But if one looks at any industry, any space of economic activity, fundamentally there would be initial years the situation would be a little sketchy and industry will try to figure out the right path. Eventually what will happen is that the stakeholders would converge at some point of time and it would become a win–win situation. Even in India, while we have had a slow and sketchy start, there is bound to be progress.

OFFSETS AS RISK-SHARING INSTRUMENTS

So what is the key to a successful offset programme? First, let us consider why companies globalize. The reason is simple: for growth, profitability, low cost and lastly for strategic alignments. As for strategic alignments, aerospace is a very capital-intensive industry and to that extent, a very high-risk industry. To develop an aircraft engine from scratch can cost anywhere between US$2 billion and 3 billion. Hence, what a company is doing is essentially betting that amount of money without knowing the final outcome. One may try and figure that out but no one knows for sure how many engines one is going to sell.

So you invest in that technology and eventually that engine does not sell or does not go onto a programme, then you are stuck with that investment. And to that extent, 'risk-sharing partners' is one particular concept that is very significant in this space. For example, GE has risk-sharing partners spread all across the world right from Japan to Europe and Latin America and beyond.

Also, at the end of the day, the concept of offsets is leverage. It has to be a win–win situation, it has to make sense for both sides otherwise it will fall flat. And that is why successful offsets really need a business plan that is viable. The tag line is that though offsets should assist good business plans, they can never be a substitute.

INDUSTRY POLICY OBJECTIVES

So what are India's industrial policy objectives broadly? It is all about creating technologies, becoming self-reliant, growing exports, becoming competitive in the marketplace and more. It is all about *Swavalamban*. What is worth knowing is the scenario of the policy on offsets in the 2005/2006 scenario when it came out.

When the policy initially came out, the six buckets shown in Table 10.1 were the big items that were worrying the industry; both domestic industry and international primes. Two particular considerations were dual use technologies and indirect offsets.

Table 10.1 Policy on offsets

Offsets policy 2005–2006 ... Industry wish list	Status
Offset credit for transfer of technology/ co-production	No progress
Indirect offsets, dual-use tech	Civil/internal security included
Offset banking and trading	Banking allowed for 2 years
Multipliers for hi-tech investments	No progress
Central agency to administer offsets	DOFA created ... more needed
Raise FDI cap from 26 to 49 per cent	Enhanced to 49 per cent

Investment in civil aviation and internal security has been allowed now. There are companies for example, like GE, which has a lot to contribute on energy. GE does a lot of high-tech work on the energy side, water and healthcare as long as needs of the national policy is served and objectives of offsets achieved. Why should the technology acquisition be limited to self-defined silos?

Multipliers and transfer of technology (ToT) co-production are also two big items. If you really want to deliver on your objectives of acquiring technology, then you need to create enough motivation to the foreign primes to bring in that technology that does not exist today.

POTENTIAL MODELS AND CHALLENGES FOR THE INDIAN PRIVATE SECTOR

Potential models for the private sector and the industry within India are worth special attention: what are these potential models, if one looks at manufacturing per se and how can we engage with the private sector in India? This is in order of complexity and growth. One could have a simple global sourcing model in which the private sector company can become a Tier-1 or Tier-2 supplier to different original equipment manufacturers across the world. The second option is of global primes bringing in their global suppliers into India and helping them set up shop. The third option is that of the risk-share programmes that would need investment upfront from the private industry and then they get revenue gains as and when the business grows. The fourth option is the joint venture, the local partner approach.

If one assesses where things are in India, the first one and the fourth one are the most talked about. So, there are a number of small-and-medium-sized enterprises and small scale companies that are trying to grow as defence and aerospace manufacturers, and approach GE on a continuous basis, claiming that they produces numerous parts. But if one really wants to be a player in that space, then one has to compete. These are the days of global supply chain where the company has to compete against Korea, Taiwan, China and Eastern Europe.

As brought out earlier, the business plan has to be viable. It cannot be run on a subsidy basis. The basic issue here is that the growth is not happening because the fundamental structure in India as of now is nascent. The country does not have the raw materials that are required, especially the exotic materials that get consumed in aerospace. Today if there is a manufacturer in India, he would have to source all that material from the Western countries, i.e. Europe or the US, make the product and then send it back. There is lack of infrastructure in terms of capacity, capability, special machines and the various requirements. The other one is that there is a need to adopt a long-term view. This industry unfortunately is a long payback industry. If you start developing an engine component in India, it would take anywhere from three to four years before that part can actually go into the engine because of the certification and safety requirements. There are multiple levels of approvals and certification that have to be obtained before an engine moves into production. So what you need to make is a call as a manufacturer in India is, 'am I going to put in five or ten million dollars today to produce a part which is only going to get commercialized four years from now?' It is a long payback game and the country needs to stress on developing the fundamental structure.

CONCLUSION

Over the years, the Defence Procurement Procedure process has streamlined and the country has a good policy that will continue to be tinkered with but at this point of time it has really arrived in that sense. The decision process has become faster than what it used to be. As a country, India took 20 years to buy the Hawk trainers. The process was started in 1984 and finished it in 2004. But in the last four or five years, if one looks at the decisions, they have happened in the two- to three-year window. Hence, the offset obligations are today piling up but the rate at which industry can fulfil them is not keeping pace. What is the way forward? In my mind, offsets should enable good business plans, it should not substitute them. The Ministry of Defence is moving forward, the Government of India is moving forward, the offset policy is evolving and it needs to keep pace with the growing offset backlog. The Indian industry as yet is not ready to absorb all the offset volume that is getting added up. The key areas that we believe can really enable the whole process as of today is to allow offset credits on ToT co-production, allow multipliers, relax foreign direct investment (FDI) policy and establish a single empowered agency.[1]

---

[1] Editors' note: ToT and multipliers have since been incorporated in the Defence Offset Policy.

11

FDI in Defence Offsets

Thelakat Jayadevan

Abstract

Thales signed a contract with India in 2009 under Defence Procurement Procedure 2006 and it had an offset part to it. So, Thales was among the pioneers in the offset area. The company has learnt a lot, and it has been an educative experience. In this chapter, the author shares some of the lessons learnt as a part of this experience.

INTRODUCTION: WHY SHOULD AN OEM BUY FROM INDIAN INDUSTRY?

Generally, when an original equipment manufacturer (OEM) prepares an offset package as a part of a bid, his objective for buying from the Indian industry (as part of offsets) is one or several of the following:

- *Utilizing some of the products of the offset package in the main contract that is to be executed:* This would be the case if one decided to buy a shelter, a power supply or something else. We could then find the use of these products in other systems and equipment that we manufacture.
- *Cost and time advantage:* Since the item is being made in India, it has a cost and time advantage.

- *Service benefits:* Finally, in any contract that an OEM signs with the Defence Ministry, a good part of the work is to be done in India. The OEM could use the capabilities of the local industry in terms of services to be included in the off-set package. In many cases, we found that the products that they were selling needed to be adapted because it did not exactly meet the requirements to begin with. So, we could try to leverage the strengths of the local industry and the local capability for development.
- *Larger strategy:* It is to use this as a part of a much larger strategy, which goes beyond offsets, for business development in India.

Irrespective of any of these objectives, the bottom line and the underlying criterion that all OEMs would use would be to have the right product or service at the right time and at the right price. This is a general idea or perspective that all OEMs usually have when they look at offsets as part of a bid that they are making. As the Defence Procurement Procedure (DPP) lays down the criteria for selection as the 'cheapest acceptable offer', the OEMs would find it much easier if the offset policy that is laid down within the same DPP would provide sufficient avenues in the area of offsets to achieve the goal.

INDIAN INDUSTRY IN OFFSETS

There is a need to understand if the part of the Indian industry that is considered eligible under existing policy on offsets is capable of meeting these objectives, which I just mentioned and would come up to the expectations of the OEMs in terms of the technology and competitiveness. Some parts of the Indian industry that seek to share a part of the offset pie may need to be bolstered and one approach could be an association with a foreign OEM. To come up therefore to the expectations of the OEMs and meet their objectives and criteria that I stated earlier, an investment, not necessarily financial, would be required. It just means that in the larger dictionary meaning of the term 'investment', the foreign OEM would really need to invest in developing the Indian industry if he wants to get a long-term advantage.

OEMs PERSPECTIVE ON FDI IN OFFSETS

A key issue is whether increasing the foreign direct investment (FDI) from 26 per cent to 49 per cent and beyond is going to help meet the expectations of OEMs. Before we discuss the figures of 26, 49, 79 and 100, we need to understand initially what might be needed to bolster the local industry. What has been experienced in the course of several bids that we have made to the Ministry of Defence is that these cover three areas: technology, competitiveness and opportunities for export. With these three issues, there could be several questions that the OEMs may wish to ask:

- Will an investment of 26 per cent allow them to shape the company in all these three dimensions? Will they be able to support technology, bring competitiveness and give them opportunity for export?
- Will 26 per cent of the profits be sufficient to bring in technology that is at par with state-of-the-art?
- Why are there restrictions only in FDI in offsets when Department of Industrial Policy and Promotion does not impose any such restrictions on services? It pertains only to manufacturing. So technically, it is felt a 100 per cent foreign-owned company should be eligible for offsets as long as it sticks to services.
- If the aim of the government is to bolster exports, why not adapt from policies that provide encouragement in several other sectors. Other sectors promote 100 per cent export-oriented industries but no such scheme is applicable in the defence sector.

OEMs would certainly welcome FDI going from 26 per cent to 49 per cent and beyond so that the expectations of the government, the industry and the OEM are met in a win–win manner that we all stand to gain out of this exercise.

CONCLUSION

In many cases, companies make an investment decision in a much larger context. There would be very few companies who decide to invest in a country just to meet certain objectives of offset. One needs to look beyond offsets as far as FDIs are concerned. Therefore, holistically, people would invest if they want to get a larger market share of defence business in India. Hence, enhancing FDI across the board would definitely bolster Indian industry at large. FDI should be looked at in a more holistic sense, and definitely if all the objectives have to be met, it needs to go beyond 26 or 49 per cent depending on the objectives the government is setting out to achieve.

12

Essential Elements of a Successful Offset Policy

George B. White

Abstract

The single overarching requirement for a successful offset policy is a collaborative relationship between the government customer and the foreign original equipment manufacturer (OEM) in the successful execution of the offset obligation. This collaborative relationship should be built around the goal of achieving 'win–win' outcomes for all the stakeholders: the government customer, the indigenous industry and the foreign OEMs. Underpinning this collaborative relationship between the government customer and foreign OEM are three essential elements for success: (1) an offset policy that is aligned with the industrial objectives it is intended to accomplish, (2) a consistent application of the offset policy across all defence acquisition programmes and (3) flexibility in the implementation of the policy in order to achieve win–win outcomes. This chapter explores each of these three essential elements for success and offers suggestions for the India Ministry of Defence to consider in further establishing these essential elements within its offset policy.

The views of Boeing Defence, Space & Security (BDS) on the 'Essential Elements of a Successful Offset Policy' are based on its considerable offset experience around the world. BDS has completed more than $33 billion of offset obligations on time or ahead of schedule, and currently has 47 active programmes valued at more than $15 billion in 17 countries representing over a dozen BDS products. Two of those active programmes are currently in India, for the P-8I and C-17.

In Boeing's experience, the single overarching requirement for a successful offset policy is a collaborative relationship between the government customer and the foreign OEM to successfully execute the offset obligation. This collaborative relationship should be built around the goal of achieving 'win–win' outcomes for all the stakeholders: the government customer, the indigenous industry and the foreign OEMs.

Underpinning this collaborative relationship between the government customer and foreign OEM are three essential elements for success:

- An offset policy that is aligned with the industrial objectives it is intended to accomplish,
- A consistent application of the offset policy across all defence acquisition programmes and
- Flexibility in the implementation of the policy in order to achieve win–win outcomes.

In the case of the first of these essential elements, proper alignment of the offset policy with the industrial objectives, it is understood that India's primary defence industrial objectives are the following:

- Creation and sustainment of high-quality defence jobs,
- Acceleration in the maturity of the defence technological base,
- Increased indigenous capability to build and support defence platforms and
- Enhanced global competitiveness of public and private sector firms of all sizes.

The key offset policy measures laid out in the Defence Procurement Procedure (DPP) for accomplishing these objectives include

- Direct purchase of eligible products and services and
- Direct foreign investment in either eligible Indian companies or research and development agencies.

In Boeing's experience, two other policy aspects play a key role in achieving desired industrial development objectives. They involve *time* and *technology*.

The aspect of *time* translates into the offset period of performance, and the longer the period of performance, the greater the opportunity for establishing long-term business relationships. Longer-term business relationships lead to more business opportunities and the creation of high-quality jobs that are more likely to be sustained once the offset period of performance is completed. Longer-term business relationships also enable higher-quality business opportunities that may take more time to establish and to pay off.

For these reasons, the majority of Boeing's offset customers around the world have offset periods of performance that extend beyond that of the main contract. Examples include the United Kingdom, Australia, Canada, the Netherlands, Korea, Switzerland and Italy.

The second policy aspect of *technology* has a direct bearing on an industry's capabilities and ability to compete. Many of Boeing's offset customers have made technology transfer, or what might otherwise be referred to as foreign direct investment, a key component of their offset programmes. Generally speaking, anything that could be considered an investment in a business entity should also be considered an investment for offset purposes. Customers such as Korea, Italy, Sweden and Finland have offset policies that recognize the value of such non-cash investments as intellectual property, know-how and capital equipment. Boeing has worked with these customers to leverage offset programmes for the development of indigenous companies and R&D agencies through training and education, modernization of equipment, technology development and research. The result has been an increase in the technology level of the indigenous industry as well as an increase in global competitiveness, as the improved capabilities of the industry have led to purchase contracts to provide goods and services to Boeing and its extended network of suppliers.

In order to apply a broader definition of investment to the foreign direct investment provisions in the DPP, a valuation methodology for non-cash investments must be developed and implemented. Boeing has such a methodology along with a documented process that is used with its other offset customers around the globe. This process consists of two primary components:

- Calculating the value of the benefit to the recipient organization and
- Determining the invested value of the transferred knowledge and capability from the OEM.

Boeing has shared this valuation methodology with the Ministry of Defence (MoD) for its future consideration.

Regarding this first essential element for a successful offset policy, one that is aligned with the industrial objectives it is intended to achieve, the following points are offered for consideration by the MoD:

- The offset period of performance should be uncoupled from that of the main contract, and should be negotiable between the MoD and the foreign OEM. This will lead to greater business opportunities for Indian industry made possible through longer-term relationships with the foreign OEMs.
- The definition of foreign direct investment should be broadened to include cash as well as non-cash investments, such as intellectual property, technology, know-how and capital equipment, with fair value recognition; and these investments should be made available to all industry sectors.

This will lead to enhanced global competitiveness for Indian industry through greater access to advanced technology, tools and processes.

One last point regarding alignment of the offset policy with the industrial objectives involves sub-contractor participation in the fulfilment of the foreign OEM's offset obligation. Limiting sub-contractor participation limits the potential business opportunities available to Indian industry, and will make it that much more challenging to achieve the ambitious industrial objectives that India has set for itself. For most OEMs today, including Boeing, 60–70 per cent of any given product's value tends to be sub-contracted, so limiting participation of sub-contractors has the potential to significantly limit business opportunities across many sectors of Indian industry. All of the countries Boeing works with around the world encourage sub-contractors to perform offset on behalf of the OEM. Consequently, it is recommended that the MoD make sub-contractor participation universal across all programmes and without the requirement for tripartite agreements, which will only increase the administrative burden and cost for the OEMs as well as the Indian offset partners.

The objective of the second essential element for a successful offset policy, consistent application of the policy across all programmes, is to ensure that all offset resources are aligned across all fronts in order to achieve the desired objectives. One way to achieve this consistency of purpose is through centralization, in other words, to have a single, centralized organization with a permanent, professional staff that is responsible for the interpretation and implementation of the policy. Referring back to the first essential element for success, this single offset organization must also have the mandate to ensure that offset policy interpretation and implementation ultimately support achievement of the government's industrial development objectives. In the absence of a single offset organization, particular care must be taken to ensure that appropriate policy interpretations and implementation guidelines are flowed out to the various responsible offices. Also, care must be taken to ensure that consistent policy implementation continues in the face of ongoing personnel turnover. If consistent guidance is not ensured, one of two things is likely to happen, neither of which is desirable:

- Either the policy may be implemented in an ad hoc fashion; or
- Inaction may set in, which could limit progress entirely.

In either case, the industrial objectives behind the policy are unlikely to be met. Most of Boeing's offset customers have established single offset authorities, including the United Kingdom, Australia, Canada, Korea, Spain, Saudi Arabia and the United Arab Emirates, to name a few.

Given India's ongoing defence recapitalization and the magnitude of forecast offset obligations, as well as India's ambitious industrial development objectives, it is recommended that the MoD consider establishing a single offset authority as the most efficient means of ensuring consistent and effective application of the offset policy over the long term.

The third essential element for success is flexibility in the implementation of the offset policy in order to achieve win–win outcomes. The reason that this particular element is important is because aerospace and defence is a very dynamic business. Platform capabilities are continuously increasing in order to meet ever-increasing requirements, and these in turn continuously drive the evolution of technology. It is within this environment that India has established its objectives to develop a more highly skilled and globally competitive indigenous industry. The more dynamic the business environment, the more beneficial it is to the indigenous industry to allow for offset programme flexibility.

Flexibility means that business opportunities may be afforded to the local industry that could not even have been envisioned at the time the offset contract was defined. Flexibility also means that the government customer may revise their policy objectives during the course of an offset programme, and still see those new objectives realized within that programme. Indeed, Boeing's experience around the world is that the more prescriptive the offset requirements, the more narrow are the benefits to the local industry.

For India, the DPP states that no changes are permitted to offset components or values after contract signature, and that changes to offset partners may be permitted only in exceptional circumstances. In order to help realize win–win outcomes for all stakeholders, including the MoD, Indian industry and the foreign OEMs, it is suggested that greater flexibility be accommodated by permitting changes in the offset contract with respect to projects, values and partners, with the written agreement of the MoD.

In summary, Boeing's experience has shown that the overarching requirement for a successful offset policy is a collaborative relationship between the government customer and the foreign OEM in the successful execution of the offset obligation. Underpinning this collaborative relationship are three essential elements:

- A policy that is aligned to the industrial objectives it is intended to accomplish,
- A consistent application of the policy across all programmes and
- Flexibility in the implementation of the policy in order to achieve win–win outcomes.

Suggestions are also provided for the MoD to consider in further establishing these three elements within its own offset policy.

In closing, Boeing commends the MoD for their willingness to solicit and positively consider changes to their offset policy. The MoD's openness to considering alternative views has resulted in an offset policy in DPP 2011 that is much improved from the initial policy of DPP 2005, and Boeing looks forward to continue engagement with the MoD in the continuing evolution of India's defence offset policy.[1]

---

[1] Editors' note: DPP 2013 had many far-reaching changes over DPP 2011. More changes are expected to follow in 2015.

13

Offset Policy: The Link between the OEM and Local Industry

Yannis Ailianos

Abstract

This chapter provides a perspective on how an offset scheme shall be best tailored to achieve national aims. The author believes that a successful offset scheme shall have inherent flexibility to take advantage of the constantly developing industrial environment. A fixed and rigid policy may lead to loss of opportunities for local industry and national security interests.

INTRODUCTION

The chapter is structured around the theoretical discussion on the offset policies. It also includes two practical case studies illustrating the theoretical discussion along with some conclusions.

OFFSET AND KEY ACTORS

The offset practice is pursued today in more than 120 countries around the globe, each with different offset approaches and rules. We also have quite a number of offset definitions. The one that is simplest to read says, 'Industrial participation/offsets are in-country compensations required by governments as a condition for purchasing major goods or services'. So offsets are essentially the conditions for sale. This implies that there are three main players in the offset environment: (a) the foreign contractor, (b) the local recipient (e.g. industry) and (c) the government requesting offset.

Any offset policy serves as a link between the three, whereby governments shape the frame of cooperation between the foreign and the local entities. The foreign contractor is to give compensation to the offset recipient country and all this functions under the umbrella of policy that is established by the governments.

ASPECT OF OFFSET POLICY DEFINITION

As the government decides to establish the offset policy, there is a sequence of interlinked events for defining such policy. First, the government defines the long-term policy objectives, the goals, the reasons for establishing an offset policy and its aspirations, the strategic fit with industrial goals, national priorities, etc. It defines long-term objectives and goals. Once this is done, typically the authorities then determine the offset parameters. Parameters such as volume, what are the eligible areas, eligible contributors, multipliers, timeframe, etc. (i.e. the 'offset guidelines' or rules) obviously these parameters need to come in support of the long-term policy objectives. After defining these parameters the government can issue tenders including the offsets requirements that the foreign contractors can respond to. But this is not the end of the job, once the tenders have come into contracts there is this very important key element that needs to be emphasized. The government establishes an efficient monitoring mechanism to ensure compliance with requirements and overall fit with goals during the implementation period. This is to make sure that the requirements have been achieved and to protect the national interest. It also acts as a support system for the dual implementation of the offset schemes to enable the long-term policy objective.

DEFINE POLICY OBJECTIVES

Governments have a large diversity in policy objectives across the globe. I have categorized those policy objectives into various categories as per global practice. Some policies focus on broader sectors, e.g. defence industry, aviation, bio-technologies and information technologies. Some objectives are focused on national priorities, e.g. security and self-reliance. Whereas some of the objectives have focus on the means to perform offset, e.g. job creation, profit generation, technology insertions, etc. Another category is the one that lays focus on activities, e.g. investments, sub-contracting, technology transfer, and maintenance, repair and overhaul (MRO), and there can be a combination of all the above elements. What is important and what we as prime contractors would like to see and we think that is very efficient in order to understand where the different nations go and what they want from us in terms of offsets is basically a clear formulation of the policy objectives which should be concrete, realistic, accounting for local recipients' capabilities and within international competitiveness serving the national policy aspirations. So today if the policy for India can be self-reliance, then this is also something that could be clearly defined in terms of policy.

DETERMINE OFFSET PARAMETERS

When it comes to determining the values of the offset parameters, we find this to be a tough job. The offset parameters need to create the environment whereby the foreign contractor will be directed to perform the activities that the government thinks are appropriate for achieving the objectives set by the policy. There are two major means at the disposal of the governments to actually achieve this goal: (1) the first means is to regulate the perimeter of activities that are eligible and acceptable for performing offset and (2) the second one is the incentivizing of the foreign contractor through offset multipliers or similar means. These incentives to the foreign contractors can basically motivate them to perform activities that the government wishes to see. Today in India we see that the policy and the means to achieve the goals focus on regulation. So it is very strong, very strict on what is allowed and what is not allowed in terms of offset.

There are some key offset-related questions that the governments and the authorities need to ask their own selves and come to some conclusions. Questions relate to the level of offset/percentage, duration of implementation, eligible parties to perform offset, eligible areas of offset contribution, eligible offset activities, non-fulfilment of regulations, etc. It is very important to know the offset value; does it have to be 50 or 35 per cent, more or less? What should be the duration of implementation? You can have low durations, high duration co-terminus to the main contract or other. Eligible participation for offset and eligible areas of offset contribution is another question. The tough job here for the government authorities is basically to take all these offset parameters, add them up and make a mix of the parameters that best support and best lead the foreign contractor to contribute to the offset policy.

ESTABLISH MONITORING MECHANISM

Another component of the establishment of an offset policy is the establishment of monitoring mechanism. It is established to keep track of the implementation, including timely grant of offset credits to contractors and steer of recovery measures where necessary. When a foreign contractor submits an offset plan to an authority this is typically a plan, it is exactly what it is—a plan. In practical terms, offset plans may prove to be implemented in practice differently than originally thought; thus, flexibility is required to accommodate for the variations in offset implementation. But this has to be monitored and also accommodated. It is the government's responsibility to keep track of the contractor's performance, grant the offset credit due to the contractor in due time. We must respect that industrial cooperation is a practical discipline and thus focus is required on the reality behind a given situation rather than on formalities. We have a case in our country where we got the first credit 12 years after the first submission of the offset credit claim. Twelve years is a long time! This also blocks the foreign contractor from taking part in more activities in the country. The government should also ensure that requirements are met and also takes suggests and accept recovery measures in some cases where required. Basically the government has to be flexible and change because change will happen.

ILLUSTRATION WITH TWO CASE STUDIES

Next, let us consider two case studies to illustrate this discussion.

Case study 1: Assume a government has the policy objective to develop the capacity and capability of the country in the sector 'XYZ'. To enhance capacity, the country needs investments, infrastructure, technology transfer, grants and training, besides the giving of incentives to industry to help them grow. Now let us take three situations pertaining to the mix of offset parameters:

(a) High multiplier for transfer of technology (ToT), high multiplier for investments and grants, and incentives to attract global workload
(b) High multipliers for ToT; however, no eligibility for investments and grants and restrictions on eligible pack of workload; only exports. Case 3; the offset parameters include no ToT, no investments and grants and no incentives to attract global workload
(c) No ToT, no investments and grants and no incentives to attract global workload

What do you think will happen? We believe that this is what will happen: In the first case, the mix of parameters will lead the foreign contractor to develop the industry as the government has actually set as objective. In the second case, ToT will come, capability will be increased but capacity will probably not be there unless the local industry also invests in increase of capacity. In the third case, we see that no matter how much the government may want to attract the foreign contractors with incentives to attract global workload; if ToT is not allowed and if investments are not allowed, eventually the growth of the industry will be minimal.

Case study 2: The second case study comes from the side of the contractor. Assuming a contractor, say Cassidian, has an obligation in three different countries and its business partners want to give its grant to one of the countries to perform offsets. Again, we have three countries with different parameters.

(a) The first one, incentive with ToT and to attract global workload but only the contractor is eligible to implement offset.
(b) The second one, valuation of offset is focused on job creation, no multipliers and eligibility errors are diversified.
(c) The third country has parameters, including offset valuation based on profits; generic and country specific and high incentives for activities in the area under focus.

Now, what will happen if we want to give this to one of the three countries under these regulations? Initially it seems as if none of these countries will get work and this is a pity. Now what we will do as a contractor is that we will go to country A, B and C and negotiate. We will say here we have something that is very special, very particular that you might want to have and try to see which government would accommodate the parameters in order to accept this work. We had a case in one country two years back where the country eventually even changed the offset rules simply to accommodate our investment, our transfer of technology and our work. For two years now, the government is very happy to have this investment that generates hundreds of millions of Euros of turnover and investments in the long term.

SOME CONCLUDING THOUGHTS

A perfect set of offset rules and guidelines does not seem to exist as rules and guidelines need to be tailored to individual govern-ments' policies and objectives and therefore to be dealt with case by case. Alignment between offset policy objectives, rules and spirit of monitoring is the key to optimized benefits to the country. The practice shows the broadest and most flexible schemes with high incentives bear the most convincing results and hence are most successful in the long term. Offset schemes need to account for the wider picture locally and globally because there are some other countries right there that will accommodate the foreign contractors' needs and will get good investments in their focus area. In total, the best practice in offset calls for close cooperation of involved actors including local authorities. Offset is a joint achievement and if it works shall be a joint success.

14

Nurturing Long-term Relationships through Banking and Trading of Offsets

R.S. Bhatia

Abstract

Although economic jurisprudence may question its relevance, defence offsets are an essential part of the business of defence sales. European countries have offsets at 100 per cent of contract value and more, while the proportion in India is 30 per cent, a very low figure. Is our loss somebody else's gain? There are three major stakeholders in defence offsets, namely the foreign original equipment manufacturer (OEM), Indian industry and the government. The only stakeholder who may pitch against offsets would be the foreign OEM, as he would like to hold on to his technology advantage. The recent enhancement of the banking provision to seven years is a step in the right direction and should be increased to 10 years. The author is, however, against the policy of trading of banking credits.

REALITY OF OFFSETS

Offsets in defence sales have been one of the most discussed subjects. We have often heard economists and some experts claim that offsets are bad. They are against the very principle of free trade and have no place in the globalized world of today. I tend to agree that these are bad. I also agree that offsets do not bring any economic value; they carry significant costs and have no place in the globalized economy of today. It is also agreed that offsets alone cannot help save a failing business and industry outcomes associated with offsets have failed to live up to their promise. In nutshell, it shows that offsets cannot leverage general economic development. It is difficult to disagree with a logical conclusion of this kind. One important result of this conclusion should have been that at least in the developed world, offsets should have been eliminated from defence sales. However, the ground reality is far from this. This brings us to another important question as to then what is the reason for such a wide prevalence of offsets. Some studies put the global average of offsets to nearly 60 per cent of all defence sales. Is there something economists and experts have missed?

EUROPEAN UNION AND OFFSETS

Let us take the example of European Union (EU).[1] The countries forming part of the EU area are mostly developed economies strongly believing in a globalized world without barriers. Going by the previously concluded logic, this should be the first Union to reject the offsets if they are so bad. It would be therefore interesting to know the model adopted by the EU for offsets in defence trade. Let us take a look at the treaty signed in 2009 by all the members of the EU less three (Romania, Denmark and one more did not sign). This agreement signed by nearly all members of EU deals with the conduct on defence offsets in the EU. It says that the primary purpose of offsets in the EU is to promote European defence technological and industrial base. It also advises the EU members to exhibit self-restraint and limit the offsets to 100 per cent of the contract value.

The obvious deduction is that members must be going in for offsets far exceeding 100 per cent of the contract value. It is only then that such an appeal can be justified. Now compare this with

---

[1] Wikipedia. EU position on defence offsets.

the ground reality in India where the offsets have been pegged at a modest 30 per cent of the contract value. The detractors of Indian policy will opine that Indian offsets are direct offsets whereas off-sets in the EU includes transfer of technology (ToT), multipliers, etc. Indian offsets have allowed this only very recently at a limited scale. However, it would be interesting to see the breakdown given in the same report regarding the offsets realized during the period under report (Ref: Wikipedia European offsets study). The report indicates that during the period of the study up to 2007, the offset value stood at about 110 per cent of the total imports by various EU members.

Interestingly, the further broad breakdown was that nearly 40 per cent of this 110 per cent was direct offsets in the programme itself and included co-production and sub-contracting and excluded multipliers, ToTs, investments, counter-trade, etc. It was pure sub-contracting and co-production. Thirty-five per cent of it was indirect military offsets and the balance was counter-trade, ToT, multipliers, etc. In the light of this are we to agree that 30 per cent offsets under Indian policy is too much and needs to be further diluted? Are we as a nation not missing something while the world is leveraging that? At the global level, it has to be a *zero-sum game.* But our present short-sighted policy is helping others benefit at our cost. Our loss is surely going to be someone else's gain.

OFFSETS AND ITS STAKEHOLDERS

Let us take a look at the three stakeholders in this whole game. There are Indian original equipment manufacturers (OEMs) too but firstly, they are a very small majority as of date, and secondly, there is no clarity in the role of Indian OEM towards offset obliga-tions in the offset policy. Let us therefore take a look at the most important stakeholder in this equation, which is a foreign OEM. It does not require extraordinary ability to realize that given a choice, all foreign OEMs would like an acquisition process devoid of offsets. In case offsets are unavoidable, then these must be as less as possible. It is therefore logical that foreign OEMs will always pitch against offsets. It is also quite clear that offsets by themselves will never bring in real technology and OEMs are not going to part with technology to merely fulfil offset obligations. India is a growing market and will spend billions of dollars in the years to come on Armed Forces modernization. The best way to stay in this market and also hope to get repeat orders would be by safeguarding the technology. So the primary question faced by the OEM would be how to safeguard his intellectual property rights and technology. The options would be to promote counter-trade or offer low technology manufacturing, or offer investments in non-strategic sectors and all of these would count towards fulfilment of offset obligations. There is no point being critical because every OEM would behave in the same manner. Added to all this is the pain associated with the disturbance of the existing supply chain and effort involved in developing new supply chain in a new country.

The second stakeholder in this is the Indian industry. Indian industry would certainly like to have greater revenue and if possible greater profit. It would look at offsets as catalysts of growth for both domestic and export markets. But, the other main concern would be the desire to get real technology and to leverage this technology for growth in both defence and commercial sectors. The pondering question would be whether the technology can give the know-why in addition to know-how. In addition to this, the Indian industry would also hope to benefit from getting exposed to manufacturing excellence and good global practices. It is relevant to look at the Brazilian model at this stage. They believed that offsets have to be treated as a partnership. Their government also made it clear that it will buy equipment from only those OEMs who will help domestic industry replicate that equipment and it was successful in doing it. Brazil was successful because they had a clear vision and mission connected with the offset policy. We also need a clear vision and mission statement regarding offsets.

OFFSETS AND THEIR CONSIDERATIONS

I had the privilege of attending a very thought-provoking session on offsets held at Observer Research Foundation where the main address was given by Prof. Jugen Brauer.[2] It was stressed by the eminent author in a very simplistic explanation that there was no need for any country to have a defence industrial base. Global free trade should take care of all defence requirements of all countries. In such a scenario, offsets have little or no role to play as offsets have invariably failed as an instrument for general economic develop-ment. This logic can hardly be faulted on economic considerations, but the fact of the matter is that world trade for defence equipment is not run purely on economic considerations alone. Technology denials have been and will always be used as an instrument of for-eign policy. It is for a nation to decide whether it wants a defence industrial base or not and whether it wants to be self-reliant or a net importer of security. Another important deduction was that offsets when implemented with clarity and focus have helped in creating local competition to foreign OEMs. It has also been clearly established that offsets have been successful in development of the indigenous arms industry. The question to be answered by us is whether we want this or not. Do we want Indian indigenous industry to be developed or not?

General economic development was never the prime consider-ation when offsets were introduced as a part of recommendations of the Kelkar Committee. The primary consideration for introduction of offsets in the Indian acquisition system was to leverage defence manufacturing base in the country and address the needs of self-reliance. The secondary consideration was to become part of the global supply chain thereby giving impetus to defence exports. When seen in this light, the debate on multipliers, ToT and other issues actually becomes redundant. But, let us get back to our aim,

---

[2] Jurgen Brauer and John Paul Dunne. 2011. 'Arms trade offsets: What do we know?' In: C.J. Coyne and R.L. Mathers (eds). *The Handbook on the Political Economy of War*. Cheltenham, UK: Edward Elgar.

our primary consideration for having offsets which is to develop defence industrial base in the country. Now, let us emulate the EU example and make it clear that 30 per cent is reserved for direct manufacturing in defence and total offsets will be much higher than 30 per cent and the balance above 30 per cent could be met through ToT, multipliers and counter-trade in other areas such as civil aviation, homeland security and investments. Any policy change that would lower the percentage of direct offsets will be counterproductive to our primary national objective.

BANKING OF OFFSETS

This is one initiative in the policy that needs to be supported fully. It is quite in line with our Indian psyche of *saving for the future*. Defence Procurement Procedure 2008 came out with this policy that allowed foreign OEMs to start accumulating offset credits even before awarding of the contract. The limitation was the time within which these credits could be utilized which was restricted to two years. I strongly support the enhancement of the time limitation to seven years in the current guidelines. We could make it even 10 years. Any OEM, who is willing to do work before the contract, must be rewarded. Chapter 5 highlights that eight proposals for banking were received and only one has been approved.[3] We must encourage banking and there was no rationale in not approving all the eight proposals.

Now if you take a look at banking per se, there are two particular important parts of banking of offsets. The first one deals with the actual carrying out of the offset work and is all about investing into the future. The second issue deals with trading of these accumulated offset credits by the OEM. Banking of offset actually serves the national aim and the actual percentage will far exceed 30 per cent if more than one OEM decided to go for banking before winning the contract. Banking is also going to bring in trust and partnership.

---

[3] See also S.N. Mishra, Chapter 5 for more detail.

It would also be a deliberate action and will bring value to offset working because it is a pre-emptive action by the OEM based on proper analysis and deliberation and would be based on natural fit rather than a compulsive partnership. Banking to me is going to be like a business arrangement with offsets being a by-product of it and not a partnership based on and around offsets. This is where the real strength of banking lies. It will also help establish a sustainable ecosystem.

Our nation can boast of frugal engineering and high-tech manufacturing. I think this is reason enough for partnership and OEMs must come forward and start this partnership. Base it around banking of offsets but let it be based on business reasons and it is sure to become a win–win situation.

It is surprising to note that not much has happened on the banking front. One reason could be lack of clarity on the procedures but the more important reason is that foreign OEMs believe that there will be a further dilution of the policy thereby making it easier to accumulate offset credits. They believe, albeit not wrongly, that it would be worthwhile to wait and watch till the policy is finalized. It is therefore important that we formalize the policy soon and then give it time to work. Frequent changes and U-turns are counterproductive. We, as a nation, have perfected the art of U-turns, but in this case U-turns and knee jerk reactions will only make the situation worse. The need of the hour is continuity and firm positive action.

TRADING OF OFFSET CREDITS

The second major issue with banking is the issue of trading. Shall we allow trading of these offset credits or not? To be fair to an OEM, trading must be allowed, but it can lead to unavoidable problems. Transparency is a major issue with offsets. It has been referred to as 'Smokes and Mirrors' by Transparency International and can become a breeding ground for corrupt practices. Acquisition with offsets is like having two contracts running concurrently—one primary contract for the equipment and secondary one for the offsets; two contracts with deep linkages—one open to public and the other hidden behind smoke screen. This picture becomes murkier and more complex when trading also gets into the frame. In order to avoid this, we need to ensure transparency and fool-proof monitoring mechanisms. Trading of offset credits therefore must be discouraged. If trading has to be allowed then the OEM with credits should be made to trade it for another programme undertaken by the same OEM in the country. This will also be an incentive for OEMs to take a long-term view of the market and the partnerships in the country.

CONCLUSION

Offsets are not a tool for general economic development and must only be directed towards development of defence industrial base. Offsets in the country today need a vision and mission statement and an action plan to make it happen. This would require strong leadership.

PART IV
VIEWS FROM THE INDIAN
PRIVATE SECTOR

15

An Overview of Indian Defence Manufacturing

V. Sumantran

Abstract

Emphasis on indigenous products and indigenous capabilities is the only way by which India can be a strong player in global manufacturing. The capability that we seek can be forced through leveraging of defence offsets. The author elaborates upon the strategy of Ashok Leyland group that has established it as a leading partner to both Indian and global organizations in the defence space and lays down three topics that need to be discussed to decide the future of growth of defence offsets in India.

INTRODUCTION

While we have seen India's evolution in terms of its economic power, in terms of its recognition as a centre of manufacturing, we need to view it as a very important global defence customer as the requirements of Indian defence forces, in order to protect the Indian economic interests in the Indian subcontinent and to project Indian ambition as a major Asian and world power needs to be reinforced with a matching military capability.

Given the current Indian defence industrial base, India has to buy weapons and military equipment to meet its needs, making her an important destination for all major military hardware giants. This provides us with the opportunity of not only creating infrastructure and getting access to the state-of-the-art military technology but also enabling this technology import to act as a catalyst for Indian defence manufacturing industry.

This elevation in India's stature brings with it significant responsibilities and obligation. We are all familiar with the range of threats that we face, many of them predictable and others unpredictable. Who would have for instance thought that in the 21st century we would have been worried about piracy in the high seas? Of course beyond those obvious obligations or responsibilities or interests, we have long-term strategic objectives.

Energy as a resource is going to change the geopolitical maps and access to energy will become a national strategic interest for any economic super power. India is fairly short on providing conventional sources of energy and its military might need enormous requirements—both conventional and non-conventional and this has led to the development of the nuclear submarines and satellites which provide the potential energy map.

India's ability to look after its security needs and project its capability not only within its boundaries but also outside of its boundaries is going to be extremely critical for India to emerge as a strategic political and economic power in the region. It is fair to say that when one aggregates the proposed defence procurement plan across the spectrum of land, sea and air systems, in the next two decades India's expenditure in the arena of defence spending and imports is probably going to be as highest in Asia apart from China and will outnumber Saudi Arabia.

This enormous spend can be leveraged by Indian large and medium enterprises to bolster the national defence manufacturing effort by using defence offsets as a vehicle to diversify and/or strengthen their presence in the Indian defence industrial base, which till date is mostly a public sector preserve.

AN OPTIMISTIC VIEW OF INDIAN MANUFACTURING

When we deal with the aspirations for India to be a credible and strong player in manufacturing and strategic capabilities, one recognizes that we have many challenges.

In order to cater to the tactical and technical requirements of operation in a vast and varied terrain, the Indian military equipment needs to rugged to operate across a very wide range of environmental conditions.

Finally, one cannot escape the fact that with all these aspirations of significant projection of power, we remain constrained by our economic capability. In order to spend, we need to earn and for that we need a sustained high GDP growth rate fuelled by the high rate of growth of the manufacturing sector to sustain the growth in the economy.

Indian diversity is a fact where we have a growing middle class and the urban demographics, which are favourable for the economic growth and on the other end a country facing the problem of extensive hunger and poverty.

As a result our policies, our capability must be couched in language and in terms and in context that recognize that we would need responsible effective solutions that are economical and financially viable. It is perhaps justified to display some guarded optimism about where we have come. As a country, when we feel the push to be self-reliant, we have typically risen to the challenge. Whether it is the fast breeder reactor, which is perhaps the most futuristic form of nuclear power generation, or whether it is the PSLV and the satellite launch capabilities or indeed even mundane infrastructure projects, we are beginning to make things happen. These, I believe, are indicators of the fact that if put our resolve to it; we are capable of generating the results we deserve. The policy itself, in terms of defence, therefore is rightly layered on certain fundamentals. The importance of looking at indigenous products and capabilities is something that has to be lauded because without having this emphasis, we will never acquire that capability.

Though the progress we are making may be slow, the point is we are making progress.

LEVERAGING OFFSETS

Offsets are critical for us, as a mechanism that so many nations have effectively translated into acquisition of local capability or local investments. With the kind of defence expenditure that is expected over the next two decades, India has a huge opportunity to translate its power as a large consumer into forcing creation of indigenous capability through defence offsets. Public–private partnerships need to emerge as important elements in the given case. Such private–public partnerships in this domain are inevitable. Public funding alone is not going to be able to mop up all the capital that we need for the broad range of capabilities and technologies that we desire. There is an element of entrepreneurship and efficiency that the private sector can add to the overall strategy in defence. In the promulgation of policies and regulation, there is a significant amount of red tape and it should be the combined endeavour of government, national institutions and industry to find ways to eliminate this red tape so that we can make quick progress towards the objectives defined.

STRATEGY OF ASHOK LEYLAND

Ashok Leyland group over the last five years has taken a very deliberate strategy to move beyond being only a truck and bus company. And this strategy has led us to strategic diversification into the Indian defence sector and global markets. Diversification is envisaged in adjacent sectors like light commercial vehicles or construction equipment and investments in future critical technologies such as system electronics, mobility system electronics, environmental technologies and IT and knowledge systems. We are very proud of our company's long association with defence. We have established ourselves as the predominant logistics provider for the Indian armed forces. There are 80,000 of our Stallion vehicles that are in service, probably the largest fleet globally other than the famed US Humvees. This is a tradition that we are very proud of and this is a tradition that we support, not just with the sale of these products but also by providing active support for these products in service at the frontlines in times of difficulty, including providing service to our forces during the Kargil conflict, right at the frontlines and ensuring extremely high uptime for those products. This has now been expanded to a global footprint and we are now actively engaged in taking these products overseas so that Indian products now provide service in other nations. Importantly, we are creating a separate area of focus, creating a separate entity so that the focus that we have on this sector is more central to our state as an industrial enterprise and this becomes a point of focus for defence-related systems.

This focus has now led us to significantly broaden our product range that goes well beyond logistics and we are moving into tactical and military application vehicles, where we leverage our strength as strong automotive engineering company with an expertise in the manufacture of defence vehicles and therefore is a very significant expansion both in our development and technology capabilities as well as the manufacturing footprint of the company.

For this purpose, we have also forged alliances and partnerships with both Indian public sector undertakings and companies as well as some of the leading defence producers in the world and this has now allowed us not only to broaden the range but also undertake private–public partnerships, development exercises for products such as the infantry combat vehicle, products that take our company well beyond the limited role of manufacturer of trucks and buses.

CONCLUSION

Defence is going to become an important element for India both for strategic reasons and for economic reasons. There are three topics that emerge as critical:

1. The difficulty of pursuing and executing an offset programme is the first. Clearly, there is difficulty from all the stakeholders' respective viewpoints. So there is one hurdle that is faced both in policy and execution of the offset policy.
2. The second issue is that of achieving a balance between the public sector and the private sector. Assuming that this happens, we are going to have offsets and we are going to have this spectrum of public and private sector players.
3. Another interesting question that comes up is: Would we in India evolve a different structure or would the tiers that are prevalent in the West also prevail here?

These are fertile topics for discussion and debate and solutions to these would decide the direction for future growth of defence offsets in India.

ACKNOWLEDGEMENT

The author would like to thank Mr Ratan Shrivastava for his assistance in preparing this document.

16

Dealing with the Government:
An SME Perspective

Arvind Lakshmikumar

Abstract

Defence offsets should not be considered to be an endgame but only as a means to build relevant technologies and strengthen domestic industrial pace. The reality of joint ventures under the umbrella of defence offsets should be considered carefully. If the domestic companies do their best, build a technology, offset will happen automatically because then the company will be competitive from a price point, it will have an understanding of the supply chain and the experience to deploy it. Foreign vendors will themselves look to make joint ventures with such companies.

INTRODUCTION

The genesis of this chapter is from my experiences in the last eight years with Defence Research & Development Organisation (DRDO), Ministry of Defence (MoD) and systems integrators in India. As a domestic original equipment manufacturer (OEM), I have a very contrarian approach to offsets. I feel that we are putting the cart before the horse: offsets are the cart and technology innovation is the horse who takes a back-seat. *Indian defence companies seem to believe that offsets are the endgame. They believe that the only reason why they should exist is because they can make money out of offsets.* The general feeling is that India has got US$30 billion in offsets, even if a company gets deals worth 1 or 2 per cent, it will become a US$5 billion company. Offsets are not an endgame at all. The actual game is building the technologies that would go to the end customer. Offsets should only be a by-product of the entire process.

APPROACHES TO DEFENCE OFFSETS

Defence offsets are aimed at strengthening domestic defence industrial pace. They do not strengthen manufacturing or technology innovation. Some of the approaches for defence offsets are:

- Joint technology development
- Contract manufacturing (build to spec)
- Contract manufacturing (build to print)
- Joint ventures (JVs)
- Licensed production
- Support and maintenance

Support and maintenance can be done easily in India and licensed production too has been practised by public sector undertakings (PSUs) for some time now. As for JVs, Indian companies and PSUs, all want to establish JVs but they want to establish them with Israeli and European vendors; that too contract manufacturing for built to print contracts. A foreign vendor comes in and says, 'I will give you design spec, I will give you everything, all you have to do is just manufacture it for me'. Domestic companies agree to manufacture as it brings in money and enhances their top line. We are a still long way away from build to spec. We are not creating specifications.

When President Obama came to India he stated that the US would remove restrictions from certain DRDO labs to ensure that technology is available to them. All rejoiced at this turn of events but if one ponders deeply, one realizes that there was no reason to exult as it made our scientists complacent. Had there been a restriction on technology transfer, our scientists would have gone to the design board and built that technology even with a time penalty of three or five years. We would have actually invested and built that technology.

Right at the top of the value chain is joint technology development. As of now, we are still not anywhere near the joint technology development stage. There are a few companies like Larsen and Toubro (L&T) that actually invested a lot of money, built a lot of technology in different spaces and have reached a level of competence where they are actually competing with a global OEM. But at the strategic electronics space, it is still operating at the periphery and a lot needs to be done.

THE REALITY OF JOINT VENTURES

Indian companies are lining up for JVs; every Indian company wants to have a teaming agreement, a JV and a partnership with OEMs. All companies seem to be able to open doors, everybody knowing the minister and his friends. *The general problem across the board is very limited understanding of the technologies involved.* There are a host of companies that have been given defence production licences for making night-vision sights, opto-electronic devices, etc. These companies do not have an idea, having never made these before. The government has given them production licences to make it with the hope that they can go back and talk to a foreign vendor and eventually manufacture it. Foreign companies are not comfortable with IP sharing and rightly so. They have spent years in research and development (R&D) and have built this technology. They would be ready to share this technology only if they see value and find Indian companies at par in core technology development. However, very few private or public companies in India actually manufacture core technologies related to any of these lists. The technology readiness level of Indian companies has much room for improvement. The government and MoD have also contributed to the issue by relying heavily on PSUs. The defence production licence should not be blindly given to the PSUs but to the companies that have built similar systems. If you want to procure an omni-directional sight for a battle tank, give a production licence to a company that has built core technology in unidirectional imaging. They may not be at par or they may be only 50–60 per cent of that of a foreign OEM, but you are at least giving a production licence to a company that has a clue about what they are building.

THE ISSUE OF GENERAL SERVICE QUALITATIVE REQUIREMENTS

The specifications issued for many of the strategic electronics components are not of the top of the line technologies. The bottom line here is that we have to approach defence and defence technology development with the objective of becoming visionaries. I still see tenders coming out. And they come out from fairly respected commands. The tenders talk about procurement of systems that are built on antiquated technology.

For example, I have seen a few tenders in the last couple of years. These commands want to procure night-vision devices for their weapons. They spec out Gen II technology that would be befitting a Vietnam era war and not a modern army. Why are not we creating specs that reflect top of the line NATO procurement? As a customer, the commands need to lay down guidelines that put strict qualification criteria around the technology.

The technology arm within the armed forces that writes general service qualitative requirement (GSQR) is not abreast with the latest technology. They are generally blindsided by local reseller partners of foreign OEMs. I read some of the GSQRs and they don't make any sense. The person writing the GSQR sees a dozen vendor data sheets and he thinks let me put together a GSQR, which has all the vendor data sheets in one sheet. There is no technology arm within the armed forces that says that they will lay down the GSQRs that are technically sound, state of the art and reasonable.

The technology development process has to be top down. After the GSQR, there should be a system design, a sub-system design and technology development and R&D should flow from that.

Armed forces need technology centres. So the Army and the Navy and the Air Force all require research labs. They all have to be staffed with people who understand the specs of the GSQR and take it down to technology domains. There has to be a systems integrator. We need to have people like L&T who work with a small-and-medium-sized enterprise (SME) and say, here is a GSQR. Build fundamental technologies to address this GSQR. Do not build technologies to make battle tanks autonomous, build technologies to improve the engine in a battle tank for instance. That is something that is the need of the hour. This has been in discussion for the last five to six years and I did not see any change since then. So offsets cannot be the fundamental purpose for existence. Indian companies have to build technology, we have to be product innovators, and we have to compete. We are not in a state where we can build a multi-role combat aircraft today. There are tons of other things that you are expected to build. You go and build those, compete them and bid on a tender, lose the tender because either you are not technically superior or you do not end up being L1 because of your overheads, then you work with the foreign vendor and facilitate their offset requirements because you have the expertise in similar systems. You cannot have companies that have not built similar systems or not have expertise to be an offset partner and expect a foreign OEM to actually comply with it.

The point is if you do your best, build a technology, offset will happen automatically because you will be competitive from a price point, you will have an understanding of the supply chain and you will have experience deploying it. Foreign vendors will automatically come to you and say be my offset partner and help me fulfil these obligations.

TECHNOLOGY PROTOTYPES AND PRODUCTION

The Army sets the GSQR at the highest level. The armed forces technology centres that wrote the GSQR work with SME clusters to develop the technology. These SME clusters are very good at building prototypes. But they are not very good at making products out of it and getting it into mass production and actually manufacturing it. So we need to facilitate partnerships between the SMEs and Indian defence manufacturing firms. Instead of having the Tatas, the Mahindras and other people going in and doing JVs with foreign vendors, let them work with the SME cluster. Allow the foreign vendor to come directly instead of coming through a JV with the Indian manufacturing firms. Let the foreign vendor and the Indian defence manufacturer compete for the tender. If the Indian company wins, they can fulfil the tender contract by licensing the technology from the SME and manufacturing the product. If the foreign vendor wins, since we have built the expertise they can fulfil the offset requirement directly with the SME cluster or fulfil it with the manufacturing agent. This would benefit the entire ecosystem.

To summarize, in the short term, our focus should be on building the base technology and facilitating adoption of this technology. We should think about offsets and other counter-trade schemes only after we have reached a certain level of technological maturity.

17

Warship Building in India

M.K. Badhwar

Abstract

The author opines that offsets can provide win–win solutions if there is a willing participation between foreign original equipment manufacturers (OEMs), Indian industry and the government. As per him, the shipbuilding sector is one of the higher employment generators for every dollar invested. The Indian industry today has more appetite for risk, has more funds to invest and can influence foreign financers to invest. The shipbuilding industry in India is expanding at a rate way above the European and American markets. As a result, foreign OEMs are keen to invest in India. This can be borne by the fact that Swedish defence major Saab has recently got into a technical partnership agreement with Pipavav Defence & Offshore Engineering Company Ltd.

INTRODUCTION

I believe that the warship sector in India has been neglected over the years. People empowered have probably not recognized its true potential in economic development of the nation. Even today just to meet our internal requirements of the Navy and the Coast Guard and to some extent the internal security, the annual workload is estimated to value more than ₹10,000 crore. A warship is a product that is the largest and the most complex category of defence equipment, as also the most expensive and most labour-oriented in manufacturing. Just to bring out few other aspects related to it, this sector has the capacity to absorb advanced technologies and processes relatively fast and if absorbed well, can eliminate the necessity of re-seeking them in future. This absorption will be relatively longer-lasting. Here whatever new methods, new technologies are available and if they are shared, the Indian technicians and experts are quite well equipped to kind of self-sustain and bring in new innovative solutions thereafter. There is also enough scope to improve our operational efficiency in this field at the present time.

POTENTIAL OF INDIAN WARSHIP INDUSTRY

Warships produced at the defence shipyards in the country have been greatly admired world over. They are comparable to the best in the world and yet have cost the government exchequer roughly half or even one-third of what the other navies may be paying to other international shipyards. Just to take an example, our present 15A class destroyers under construction at Mazgaon dock roughly cost around ₹3,500 crore a piece, whereas Australian Navy has placed order for three similar-sized ships on a European shipyard at almost ₹11,000 crore a piece. That is indicative of economic gains of warship building activity in the country. This sector badly needs infusion of skilled manpower. We have not got adequate number of engineers or rather any type of shipbuilding professionals. Only the Indian Navy has been trying to push some amount of HR activity in this direction. It is the only Navy in the world which designs its platforms in house. But the combined capacity of the Defence Public Sector Undertakings (DPSUs) is not even one-third of what the Navy alone requires. And therefore the Navy is always under pressure to look beyond the Indian horizon for acquisition of warships.

The warship-building sector is also one of the highest employment generators per dollar invested. Commercial shipbuilding itself generates more jobs than any other sector but warship building perhaps can generate up to 10 times more employment compared to commercial shipbuilding. Investment in this sector can also provide handsome dividends to even the foreign original equipment manufacturer (OEMs) through enhanced indirect market share into Indian orders as well as export out of India. Significant export potential exists for warships. These are today supplied by European countries to the Third World countries. But their cost of production of these vessels in this platform is much higher than what Indian shipyards can deliver at. It is not without reason that the foreign OEMs are very keen to collaborate with Indian shipyards, particularly with the private sector ones. But even if the government wishes to keep the DPSUs in the loop, models can be evolved where both private and the DPSU shipyards can participate in such collaborations, as is being currently being envisaged through the forthcoming LPD project for the Indian Navy.

IMPLEMENTATION OF OFFSETS

Implementation and management of offsets particularly when we talk of win–win solutions has to have a willing participation of all the three partners, that is, the foreign OEMs, the Indian industrial partners and the Government of India. All three must see it as a win–win situation, only then will we get tangible results. Perhaps this is one of the reasons that we have not had many success stories thus far. In the Indian context, warship building has been restricted only to the DPSU shipyards in the past and perhaps for very valid reasons, like listed below, because of which the private sector was just not willing to come forward.

- Budgetary constraints resulting in relatively low allocation for defence acquisitions
- Very high financial outlay
- Long gestation periods
- Risk averseness of private industry

The situation has now changed and in recent times private business houses have become more willing to take risk, they have funds to invest, they are even able to influence external financiers to make additional funds available, the budgetary constraints are no longer as serious as they were in the 1990s, the relevance of the Indian armed forces requires them to augment their fleet and the foreign OEMs are becoming increasingly willing to partner with the Indian industry because their own domestic markets are sharply shrinking. As a result, more and more capital-intensive state-of-the-art world-class shipyards are now being set up. Pipavav Defence, Larsen & Toubro, ABG Shipyards and Bharti Shipyard have all set up state-of-the-art greenfield shipyards in different parts of the country. The Government of India and the Indian Navy must guide these yards and ensure that defence offsets in terms of technology, skill development and co-production of ships are made available to them through a well-designed intervention and not left to themselves to manage these affairs. It will be a pity, if the net productive capacity of these modern yards is not gainfully utilized by the nation and they will die a natural death in foreseeable future.

CONCLUSION

Gross capacity of DPSU shipyards is grossly inadequate to meet even the national requirement for warships. Therefore, we must bring in the private shipbuilding industry into the fold. Many new private players have come forward to take up the challenge. However, there is acute shortage of skilled manpower, technology and expertise in the country for optimal functioning of these yards. Indian shipyards wanting to engage in warship construction are therefore naturally preferred candidates for utilizing defence offsets. They are not only in a position to absorb the kind of offsets that are available, but badly need such offsets to supplement their assets in terms of technology, skilled personnel and modern processes. Shipbuilding is a highly capital-intensive business. It involves long gestation period, very large capital outlay in terms of thousands of crores of rupees. It is certainly difficult for them to sustain financially for long periods without financial help. Here too defence offsets can come in handy by allowing foreign OEMs to participate financially in working of these yards in terms of equity, for which Government may consider higher foreign direct investment (FDI) in this sector. It is better to raise FDI limit than to resort to import of warships, because it will help the Indian industry considerably in employment generation and the country in self-reliance. Similarly, we must use the concept of offset multiplier. Education and training particularly relating to shipbuilding professionals is grossly inadequate. Training and skill development is another area that can gainfully utilize offsets gainfully for the foreign OEMs and the Indian industry. In conclusion, the government must encourage, regulate and steer the offset utilization for the warship building industry in a big way, because this sector can perhaps provide highest returns on offset investment in terms of gearing up of the nation towards self-reliance, which is the basic objective of defence offsets.

18

SMEs as Prime Contractors

Ashok Atluri

Abstract

The author provides his perspective on experiences in offsets from the small and medium enterprise (SME) point of view. His experiences as a prime contractor and as offset partner to global original equipment manufacturers (OEMs) are valuable and raise some pertinent issues on behalf of SMEs. He strongly believes that SMEs should pitch for the opportunity of being prime contractors because there are advantages inherent to all in doing so.

INTRODUCTION

This chapter explores the option of being a prime contractor for SME and why I am recommending that SMEs should also look at being prime contractors. Zen was incorporated in the year 1993. We have been focusing on training and simulation for the past 19 years. And our focus has been on two customer segments: the police and the defence. And this focus has led us to build unique competencies in the field of simulators. We have been the prime contractor to the army since 2007; we have participated in various procurements and responded to many request for proposals. So, when we look at the offset process as an alternative we find the process long and tortuous: you search for potential prime contractors; they come, visit multiple times and they access your financial statement, do costing, and finally give you a cost plus business. We have been interacting with a few vendors to become an offset manager/offset provider but again our experience has been a bit mixed. However, we have not generated any business by being an offset partner so the view presented here should be taken as one-sided.

THE RISKS OF BEING OFFSET PROVIDERS TO GLOBAL PARTNERS

In a joint venture (JV) formed due to offset obligations, an important question that we need to ask is: what happens to the JV after the local firm meets the offset obligations? I had an experience with a friend who is from Turkey. As part of the offset programme, they were visited by the prime contractor and they built a lot of capability, invested a few million dollars hoping that they will be the global hub for all future supplies. When the OEM got the next order, they never got in touch with the Turkish company. When the company went back to the prime contractor to clarify why, the response from the OEM was that the country to which the prime contractor was supplying the new product was insisting on an offset programme— so they had to accommodate a local manufacturer. This is one of the risks you may run when you build capability assuming that you will be the global partner. You need to take assurances from your partner that your interest will be protected for the long term. I definitely think that there is lot of scope in the offset business and there will be a lot of success stories in the short term. But we need to make the success sustainable—and we need to include enough clauses into the agreement to protect our interests. The persons who are assuring you now verbally may not be around later.

BEING A PRIME CONTRACTOR

The prime contractor interacts with the customer directly. The prime contractor is responsible for delivery, for the whole execution of the contract and for after-sales support. So you really need to build a competent organization. We have gone through this process and we feel that there is a lot of value that can be captured if you are the prime contractor. So our thinking is completely oriented towards the end-user—product-focused thinking with long-term reliability and robustness built right from the design stage. The trials that the Indian Armed Forces do are the toughest by a large margin. And they are the most rigorous—this is based on the feedback that we have got from foreign competitors who have participated in various trials throughout the world. One virtue that will be helpful while dealing with this segment is patience. Patience will see you through difficult times. It may be three or four years before you get any order but you have to sustain. It helps to be low cost and efficient. Unless you are low cost and very efficient you will not be able to sustain the lean periods.

CHALLENGES AS A PRIME CONTRACTOR

There are very specific challenges and advantages that an SME may face as prime contractor:

- *The no-cost no-commitment participation:* The company is not paid for the whole process. So the company has to build a product upfront. The typical time frame to build a product would be up to 2 years and would require a large invest-ment. And this has to be self-financed. This is specifically for the 'Buy Indian' category. And there is no commitment that even after you have gone through the whole process and you have been shortlisted, the purchase will automati-cally happen. If there is a procedural flaw in the process of evaluation the tender can be cancelled. Even though the cancellations are done due to valid reasons and after great deliberations internally by the forces and Ministry—it can be quite damaging to SMEs.
- *The L1 process:* This is the biggest benefit to the Indian SMEs. Once you qualify technically and your financial bids

are open what they do is they compare you with the other bidders and if you are lowest bidder (L1), they place the order with you.

CONCLUSION

We have personally had good experience with bidding as a prime contractor. We think we have built extremely good capability within the company. We have a range of 18 simulators now. SMEs should look at the defence requirements (which are regularly circulated through industrial associations, various seminars/symposiums and on the defence website), look at their own capabilities and see what they can offer as a complete solution to the armed forces.

19

The Market Opportunity

Vivek Lall

Abstract

The author reiterates the importance of India as a huge market for defence products, especially in the phase when global defence budgets are shrinking. He is of the opinion that it is imperative for all companies intending to do business in India to adopt a collaborative rather than a buyer–seller approach. He also talks about the role of small enterprises in establishing the defence production of any country and feels India has the potential of becoming a part of the defence 'Global Factory'.

A strong, technologically vibrant and financially success-ful defence industry is an aspiration for any country. Manufacturing of defence products is something of great interest both to the public and private industry. Projects in the defence sector represent some of the best long-term growth opportunities for the industry. Beyond the business benefits of participation, manufacturing for defence, including aerospace platforms, represents opportunities for technology access and development, as well as the establishment of long-term relationships. These are critical as our industry competes for future business in an increasingly globalized market.

From a foreign company's perspective, India clearly represents a huge market opportunity. In the boardrooms of companies abroad, as they look at India on the world map, they see a tremendous market for their products, especially at a time when defence budgets around the world are shrinking. However, there has been a paradigm shift in the way companies—whether present in India or looking at a presence in India—think about the country. Increasingly, global companies are looking at the merits of a collaborative model rather than a purely a buyer–seller relationship. In that respect, a collaborative model means that different elements of manufacturing a product are in some sort of collaboration with the local industry: whether it is the R&D; the preliminary design of a product; the final design of a product; the prototyping; manufacturing; production; the actual sale of the product; and the after-sales support.

As we move into the future, it would be important for companies looking to do business in India to address each of these areas of collaboration. A strategic entry into India, therefore, would not simply mean a buyer–seller relationship but the ways in which a company collaborates with the government and the Indian industry at large. Such type of collaboration for an established large player or even established small players abroad would necessarily mean an increase in risk from their perspectives because India is a new country, a new market, a new way of doing business. But surely, for the increase in risk, the reward is huge.

The Indian domestic industry stands at the threshold of a huge market opportunity. The domestic industry, whether public or private, needs to clearly articulate its vision and mission for defence manufacturing. Of course, offsets is a great catalyst for transfer of technology and economic growth. The Government of India and the Ministry of Defence (MoD) need to be commended for the robust and an evolving policy that has been put in place. Only when one compares offset policies of other countries to what has been created in India, does one begin to comprehend the large complexity of this subject and as to how much progress the MoD has indeed made in terms of setting up a framework and guidelines. It is now up to the industry to make the best of it.

By its very nature, the defence sector requires large, long-term capital investment. Investors, therefore, cannot enter this sector with expectations of seeing returns in a year or two but have to start thinking more strategically in terms of a 5- or 10-year horizon. For the Indian industry, taking a long-term view means taking risk. But again, taking measured risks with a considered risk mitigation strategy will lead to very significant rewards in the years to come.

Defence industry is a multi-tiered ecosystem wherein you have Tier-1, Tier-2 and Tier-3 players that feed into the large system integrator. The robustness of a country's defence industry is established not because of one of two or three or four big players—it has happened because of a very strong small and medium enterprise (SME) network. If we look at the economies where the defence industry is well established, besides the prime contractors we see Tier-1, 2 and 3 suppliers and niche product manufacturers. These smaller businesses are widely recognized as being the powerhouse of the defence and security community, driving through change by being innovative and developing new technologies. For developing India's defence industry, the incubation of the SME sector—which is of critical importance to the supply chain—is very important. The SMEs are the pillars that can stand up a true defence and aerospace industry in any country.

Industry in India can truly showcase its manufacturing prowess and become a centre for excellence in manufacturing and I do believe that the right ingredients to make this target achievable already exist. Almost all major defence companies are system integrators, and have spread their various functions—manufacturing, commercial and management over a number of locations globally—chosen on the basis of relative advantages they offer. Just as India is becoming the global manufacturing hub for companies in sectors like automotive and telecom equipment, it has the potential to become a part of the defence 'Global Factory'. The time has come for India to leverage on its strengths—its manufacturing and engineering capabilities; its pool of skilled expertise; and its sheer size that offers it a strategic advantage for servicing the entire Asia-Pacific market.

Acquiring depth in manufacturing is crucial from the standpoint of long-term competitiveness in strategic areas of economy such as defence. From the perspective of national security, it is important to have a strong indigenous value chain addition element. An offset is just a catalyst. Ultimately, our aim should be to develop India as a centre of excellence for manufacturing regardless of offsets because it is the most competitive place to manufacture products.

PART V
CONCLUSION—THE LARGER CONTEXT
OF INDIAN MANUFACTURING

20

Building India's Defence Industrial Base

R.C. Bhargava

Abstract

The author talks about the various barriers to manufacturing in India, including shortage of skills, infrastructure weaknesses, limited management resources and the limitations to the development of technology in India. The success of manufacturing in India will depend on the policies adopted by the States that will see value when performance rather than incumbency and caste issues would dominate elections in the States. Increasing the foreign direct investment in defence offsets from the current 26 per cent would attract foreign manufacturers to bring their technology and manufacturing capabilities to India.

INTRODUCTION

One of the key policy objectives of any government has to be ensuring national security. In India, from the time of Independence, we too have always endeavoured to achieve this objective. Countries achieve their objective of national security in many ways. For example, having political and defence alliances is one such way. However, it is also necessary to build strong defence capabilities, which become a deterrent to any proposed hostile actions from another country. Defence capabilities require not only well-trained and equipped armed forces, but also the industrial and infrastructure base to ensure that the availability of weapons and equipment is comparable to those of the perceived hostile countries. The production of these weapons in India would ensure greater national security, than being dependent on foreign countries for their supply and maintenance. It is not unknown for foreign countries to curtail supply of arms and components, in times of need, to exert pressure on a country. In addition, having a well-developed domestic production industry for weapons enhances a country's international standing and can be used as an effective tool of foreign policy.

In the last 25 years, the world has seen the rapid growth of technology in virtually all walks of life. Often, the technology for civil use has emanated from work done to enhance defence capabilities. As a result of research and development (R&D) in the defence area, weapon systems have become far more technology intensive than ever before, and the use of electronics, both in terms of enhancing defence capabilities and also retaliating against aggression, has grown exponentially. Materials for use in defence products have become lighter and stronger. All this has called for a huge expenditure and capability building in the area of research, development and testing. Countries that have led the way in developing new technologies for defence have not only gained an edge over others but have also developed their industrial manufacturing capabilities for defence equipment to become large exporters. Selling defence equipment usually gives a political advantage. India has also, within its means, tried to keep up with the rest of the world. However, R&D expenditure has been around 2 per cent of domestic sales and there has been limited success in designing and producing new weapons. As a result, over the years we have increasingly become dependent on the import of weapons and systems from other countries. While the Defence Public Sector Undertakings (DPSUs) and the Ordnance Factory Board units have been increasing production steadily, the bulk of the sophisticated equipment, for all branches of the armed forces, is largely imported.

The export of defence products from India has been limited both by policy and the capability of meeting the needs of foreign countries. The extent of exports, in the last 3 years, has averaged less than ₹300 crore. This limits the earning of scarce foreign currency and slows down the growth of manufacturing activity and creation of employment opportunities. The ability to increase India's influence in smaller developing countries in our neighbourhood is limited by our inability to sell required weapons to them. On the other hand, China has become a large exporter of weapons and this is used as a tool of foreign policy. It is a development that should be viewed with concern by all connected with national security.

The need to increase our capabilities in the area of defence production has been realized for some time. This led to the private sector being allowed to enter defence production in 2001. Foreign direct investment (FDI) up to 26 per cent was also allowed, in the expectation that this would lead to joint ventures with Indian companies and the transfer of technology to them. So far, the progress in this direction has been, not surprisingly, slow. The government also brought in the offset policy in 2005. The policy does create, in theory, huge opportunities for the growth of ancillary industries and the development of micro-small-and-medium-sized enterprises (MSMEs). To some extent, this is happening. But here again the problem usually relates to the technological and financial capabilities of the Indian companies to produce the kind of inputs required for the imported sophisticated weapons and systems. The inability to produce parts and assemblies also restricts the scope for Indian companies taking on the specialized work of maintaining equipment after they have been bought. In many areas, our dependency on foreign companies for maintenance is high, and the cost is considerable.

MANUFACTURING IN INDIA

Manufacturing activity has not progressed as rapidly in India as it has in many other developing countries, some of which have joined the ranks of the developed world. The share of manufacturing in the GDP has remained virtually constant, at about 15–16 per cent, over a long time. Policy makers in India do not seem to have given priority to creating the requirements of achieving high growth of manufacturing. Perhaps that would have conflicted with the other objectives of self-reliance and import substitution. The high rates of growth of GDP, which we have achieved in recent years, are largely the result of the service sector growing at a high rate. On the other hand, countries such as China, Japan, Korea, Singapore and Thailand have achieved their high GDP levels with the growth of their manufacturing sectors, which even now contribute between 30 per cent and 40 per cent of GDP in China, Thailand and Singapore. It is hard to see a country as large as India, and with a population of 1.2 billion, having sustained growth without a robust manufacturing sector.

India is considered to be a low-cost manufacturing country. There is an abundance of labour and our education system, at the time of Independence, was ahead of almost all the underdeveloped countries at that time. We also had reasonably good infrastructure in 1947. With these advantages we should have been able to get ahead of other developing countries in becoming a manufacturing hub. However, for various reasons this did not happen. Our manufactured products gradually became obsolete in terms of technology and quality. We are aware of the fact that till 1991 we had virtually no manufactured products exported out of India. Our exports had fallen to virtually miniscule levels and we had only a few days of foreign exchange reserves left at that time. That is when the reform processes started, and subsequent to that reform process, a large part of the protective systems for industry, which had been built after 1950, were dismantled. We became more open to foreign investments and technology, and several industries became competitive. In areas where there is competition, our products are near world class today. The events of the last 20 years clearly bring out the importance of an open door policy for investments and technology, and encouraging competition, which leads to better quality and lower prices for the consumer.

Let us look at what has happened in the area of defence production. Being an area critical for national security, the government rightly decided that all investments here should be in the public sector. Over time, we have built a sizeable PSU base for the defence sector. A separate Ministry of Defence Production looks after these PSUs. The Defence Research & Development Organisation (DRDO) was also created to do R&D work and develop technology for the defence sector. Other areas relevant to national security such as nuclear energy and space research have their own organizations. Despite the importance given to defence production capability building, the results have not been adequate, and we have fallen behind China and other countries. Realizing the need for more capital and management resources for defence production, the private sector was allowed to enter defence production a few years ago. FDI was allowed but only to the extent of 26 per cent. It was hoped that this would result in filling some of the technology gaps.

So far this policy has worked to a very limited extent. It seems that the restriction of 26 per cent FDI is hampering the inflow of technology from abroad. Those who have developed sophisticated technology are reluctant to part with it, unless they have control over the company acquiring the technology. A reputed manufacturer of defence equipment would like to ensure the security of the technology and prevent its leakage to others. Such manufacturers would also like to ensure that the technology is used so that the products are up to their standards of quality and performance. Delivery schedules should be met and this would require that the supply chain has to be developed so that it can meet the quality and delivery commitments of the original equipment manufacturer. Our experience with Maruti has shown how hard it is to do this. These reasons have resulted in a slow inflow of technology and the import of weapons and systems continues to remain high and there is little evidence of Indian capabilities increasing significantly in this sector.

Presently about 70 per cent of defence equipment is imported. It seems quite anomalous that we should be willing to buy strategically important weapons and systems from foreign manufacturing companies, situated abroad and not allow the same companies to invest more than 26 per cent in India and manufacture their equipment in our country. Surely national security would be better met if a foreign-owned company manufactured the equipment in India, compared to importing the same equipment. The logic that does not allow this to happen can perhaps only be explained in terms of a perceived threat to the DPSUs, the unions of the PSUs and the vote banks associated with these unions. There has been a considerable body of opinion in favour of enhancing this limit of 26 per cent but so far the government has not made any change.

FUTURE GROWTH OF MANUFACTURING

The necessity of the manufacturing growth to increase at a much faster rate has now been recognized while formulating the 12th Plan. The cabinet has approved the national manufacturing policy. This requires that by the year 2022, the share of manufacturing should go from its present 15 or 16 per cent to 25 per cent of GDP. It is a very large increase and to achieve it would require that the manufacturing sector grow 3–4 per cent faster than the GDP. In the past decades, the two have been growing at about the same rate, and so a substantial increase in manufacturing activity would be needed to achieve this higher growth rate. It is significant that in the last few months the rate of growth of the manufacturing sector has dropped to 3–4 per cent and the task for the 12th Plan has become that much harder.

The policy has identified the main actions to be taken that would make the manufacturing policy a reality. It is stated that infrastructure investments have to become larger and implementation of projects more efficient. Given the state of government finances, it is obvious that for investments in the infrastructure sector to increase significantly, private capital has to be attracted in a big way. The involvement of the private sector in implementing projects is also likely to increase the timely execution of projects and their efficient operation. Policies for involving the private sector in infrastructure projects have so far not been very successful and there is no indication that the root causes of this have been identified and corrective actions taken. The private sector would invest money in infrastructure if it leads to a return on investment comparable to other alternative sectors. There is often a reluctance to accept that the private sector should make that kind of profits and there is a fear that if allowed, it could lead to criticism from other political parties, media and so on. Land acquisition and environmental clearances are two other factors, which often cause delay in projects, and these still remain very difficult areas to negotiate. The government has earlier taken a decision on allotment of government land, which could lead to one bottleneck being removed to some extent.

In 1991, one of the reasons for opening the doors to FDI was that it would result in bringing in both product and manufacturing technology. It was also stated that foreign investors would also ensure greater market access for our manufactured products. To a large extent this happened. The Planning Commission recognizes the need for more technology inflow if the higher rate of growth of manufacturing is to be achieved. Areas where technology is needed are quite well known. The question is: when will technology start coming in? In some sectors, the inflow of technology is related to the restrictions on FDI in those sectors. The defence sector is perhaps the area where the inflow of capital could lead to a large increase in manufacturing activity, including the growth of small and medium industries and the generation of employment. Another area is the processing, storage and marketing of agricultural products. It is necessary to look at the total chain of activities, from the growing of the product to its retailing, and to make a realistic plan that would encourage both Indian and foreign capital to enter these activities. Banking and insurance are two other areas where foreign capital would help support the growth of industry.

The Planning Commission has rightly highlighted the need for skill development. It is recognized that a faster growth of the manufacturing sector is not possible unless there is a larger availability of skilled workers. The question that we have to ask is: why we are lacking in skill development? We have a large number of industrial training institutes (ITIs) and polytechnics but the general experience of industry is that these do not impart the skills needed by them. There are issues regarding the motivation and quality of the teachers in the government-run ITIs. The infrastructure in most ITIs has not been well maintained and the equipment for training is largely obsolete or not in working order. Recognizing these shortcomings, the public–private partnership model has also been accepted for the ITIs. The National Skill Development Corporation has been established to upgrade and enhance skill development. Industry is being involved in modifying the syllabus and the training programmes. However, the scale and extent of involvement of the private sector is still inadequate. It is unlikely that during the 12th Plan there would be any significant improvement in the availability of skilled workmen.

It is important for industrial enterprises to employ skilled workmen. A very large number of the total Indian workforce are in the unorganized sector, and most MSMEs are reluctant to pay the higher wages associated with skilled workmen. They prefer to employ untrained people at lower wages. Several industries in the organized sector also do the same thing. As an outcome of this attitude, students would be reluctant to pay the higher cost of a good education at an ITI or a polytechnic. Skilled workmen can produce better quality of products and increase productivity. However, unless there is competition and the possibility of sick units going bankrupt, companies are not likely to appreciate the value of using skilled manpower and improving their competitiveness.

Manufacturing activity is not only experiencing shortage of skilled workers but also other hiring-related issues. It is perhaps even harder to hire good engineers, business management graduates and even ordinary graduates. At the same time, it is also a fact there are numerous individuals with degrees, including engineering degrees, who are without jobs appropriate to their qualifications. The same holds true of persons having an MBA. The reality is that the number of persons who pass out of colleges and business schools are more than adequate to meet the needs of industry. For example, we produce about a quarter of the total MBAs that are produced in the entire world, yet we are short of managers.

The reason appears to be that a large majority of the business schools are unable to impart the kind of education expected of them by the users of their products. The smaller numbers of MBA graduates who are produced by the good schools find lucrative employment opportunities in the financial, consulting and IT sectors. The manufacturing sector usually does not attract the products from the IIMs and similar good institutions. Unless the output of well-educated MBA students increases substantially, the manufacturing sector will continue to face a shortage.

We produce the second largest number of engineers in the world but the bulk of them are not found fit for employment. The reason is the quality of the education imparted in many of the colleges is such that the user industries do not find the graduates suitable for employment. The shortage of well-educated engineers results in the output from the good engineering colleges going to the IIMs and other good business schools. Most of them do not come to industry or to the manufacturing sector. Our experience is that we have to look for engineers from the second level of colleges.

With the Indian economy doing reasonably well, the shortage of managers and engineers results in high attrition rates and rapidly rising wages. This not only affects the competitiveness of our manufacturing sector, but also causes industrial unrest. Unions often point out the growing disparity between their salary increases and those of the management cadres. One of the main reasons for this situation is that the general standard of education in India has not kept pace with the changing times. The problem starts from the primary education, where the bulk of the schools are government-run. Good schools, most of which are in the private sector, are inadequate to meet the demand. Private investment in schools and education is restricted because schools cannot be run for profit. I often wonder how the public good is served by this policy that results in an inadequate number of schools that can offer good education. The lack of good education at the school stage reflects in the quality of entrants to the general and technical educational institutions. It is not surprising that much of the output from these institutions does not come up to the requirements of industry.

Another result of the small number of well-educated persons is that they choose careers that offer the best prospects in terms of remuneration and career growth. The teaching profession does not compete well with the other options available in our economy today. Thus the bulk of the entrants to the teaching profession are not from amongst the best that graduate from colleges. This is one of the reasons why the quality of teaching in a large number of institutions is below what is needed.

The growth of the manufacturing sector is thus likely to be impeded by the low quality of human resources. Industry would have to devote increasing time and resources in training its personnel and may be discouraged from doing this if attrition rates continue to be high. In any case this would be an additional cost and will impact competitiveness. The comparatively higher cost of doing business in India, a fact which is well recognized, already has an adverse effect on her competitive ability.

THE DEFENCE SECTOR

The use of sophisticated electronics in defence equipment has already been mentioned. For various reasons we do not have a well-developed electronics industry in India. Even in the automobile sector, the import content by vendors is still quite high because it is not possible to procure electronic components in India. While the use of electronics in cars has been growing, it is far less than what is required in the defence sector. It stands to reason that if India is to build its defence industry, it has to acquire the technology and the capability of manufacturing highly sophisticated electronic systems. I do not think that given the state of this industry in India, we can build the capabilities by ourselves in any reasonable period of time. The only option that can help us grow our defence production capabilities is to attract foreign manufacturers to bring their technology and manufacturing capabilities to India. The policies followed so far have not resulted in this happening. We need to review the existing policies and introduce measures that will make

India attractive for foreign manufacturers of electronic systems. Certainly a far more liberal FDI policy would be needed.

Electronic system manufacturers would probably only come to India if the users of these systems set up production bases in this country. It is most unlikely that electronic systems would be manufactured in India only for export. We do not appear to have any well-accepted advantage in costs or manpower that would attract manufacturing to shift to India. Here again, past experience shows that without a change in the FDI rules, the major manufacturers of weapons are unlikely to establish production bases in India. These holders of high technology are not likely to part with their know-how to companies that are controlled by others. Thus, the FDI limit at 26 per cent is likely to prove an insurmountable hurdle for sophisticated weapons to be manufactured in India.

GOOD GOVERNANCE AND EASE OF DOING BUSINESS

There exist various barriers to manufacturing in India, including shortage of skills, infrastructure weaknesses, management resources being limited and the limitations to the development of technology in India. Another question is how easy is it to do business in India, compared to alternative locations, and how does India compare with other countries in terms of costs of doing business? How do foreign companies regard India as a destination to establish a manufacturing business?

The World Bank has rated India as 132 in terms of ease of doing business. We could argue about the criteria used to come to this ranking and not agree with it, but the fact is that this ranking will be considered by foreign manufacturers while looking at India as an alternative to other places. The recent media publicity regarding scams and corruption, infrastructure failures, delays in decision-making and the postponement of many reforms, the Vodafone case and labour problems in India have not enhanced our attractiveness as an investment destination. What most investors do not know is that most decisions relating to manufacturing are now taken at the level of the States, and there is a wide difference between States in the facilities offered for the establishment, and operation of industries. Some States are much more investor-friendly than others. It is therefore advisable for potential investors to look at the more investor-friendly States and not go by the media reports that tend to treat India as a single administrative unit for investment.

The defence production sector, however, is one where the central government has a major role to play, especially for foreign investments. Approvals to foreign investments have to come from the Centre. The DPSUs are under the control of the central government. The Centre also controls the buying of the products from the defence production industries. Once the Centre has given the necessary approvals, the States will come into play in terms of acquisition of land, providing the necessary infrastructure and creating an environment conducive to efficient production.

Thus, the growth of general manufacturing activity will largely depend on how the State governments understand the requirements of this sector and give priority to providing the facilities for enabling industries to work competitively. If States see that the growth of industrial activity and the associated employment opportunities lead to political gains, they will do what is needed. In recent years, the elections have shown that States where there is good growth of GDP, and standards of governance are relatively high, have usually not had the anti-incumbency factor coming into play. The young voter, whose percentage is steadily rising, seems to prefer this option than be influenced by caste and such reasons.

However, for India to build its defence production capabilities and reduce the dependence on imports, the existing policies will need to be reviewed and changed. They have not delivered the results as was hoped for. Our gap in this area with China is growing. It is imperative, in the interests of national security, to reduce this gap. This can only be done if the central government takes the needed actions.

About the Editors and Contributors

EDITORS

ManMohan S. Sodhi is Professor in Operations and Supply Chain Management at Cass Business School, City University London. He did his PhD at the University of California, Los Angeles, and his undergraduate at the Indian Institute of Technology, Delhi. From June 2011 to September 2013, he was Founding Executive Director of the Munjal Institute for Global Manufacturing and Visiting Faculty at the Indian School of Business.

Rajiv Bhargava (Retd. Colonel) is an alumnus of National Defence Academy, Defence Services Staff College and the Indian School of Business. With an extensive experience in domain of Air Defence with the Army, he has keen interests in aerospace and defence; especially defence offsets, and has been instrumental in organising the 'Swavalamban' series of seminars on 'Defence Procurement'. He currently works as an Associate Director with the Munjal Institute for Global Manufacturing at ISB.

CONTRIBUTORS

Yannis Ailianos is Head of Industrial Development within the Sales organization of Cassidian, the defence and security division of EADS and is responsible for industrial and offset-related activities in the frame of major export sales of Cassidian. He has been involved in offset activities for more than 12 years, during which he has covered a large variety of offset cases in campaigns as well as in implementation. He has been engaged in more than 10 countries including Greece, India and Norway, and has worked on some of the largest offset deals for Cassidian, related in particular to the sales of military aircraft. The partnership nature of the latter has additionally enhanced his deep-routed experience in industrial partnerships. Besides his position within Cassidian, he is currently also serving as Chairman of the training committee of the Global Offset and Countertrade Association. Before joining EADS, he worked in various positions in the area of financial services. He has a master degree in Management of Technologies and a Commercial Engineering degree from the Brussels-based Solvay Business School.

Ashok Atluri is the Managing Director of Zen Technologies Limited. He has been associated with the Company since its inception in 1993. He received various awards including 'Small Scale Entrepreneur of the Year' award from Hyderabad Management Association. Mr Atluri has presented various papers, including 'Use of Simulators in Training Police'. Under his able leadership, Zen has got the highest CRISIL rating of SME1, has attained CMMi Level 3, won the National Technology Award for 2012. On a personal side, he is an avid reader with an eclectic taste.

M.K. Badhwar (Retd. Rear Admiral) is one of the highly qualified shipbuilding professionals in the country with multifarious qualifications, all from institutions of global repute. These include BTech ME (Hons) & DIIT (Naval Const) from IIT Kharagpur, MSc (Warship Design & Const) from erstwhile USSR Naval Academy, St. Petersburg, and MBA (Fin) from FMS, Delhi University. Commissioned in October 1972, Admiral Badhwar served the Indian Navy for almost four decades before retiring in December 2009. He is a recipient of prestigious national awards AVSM and VSM for the distinguished services rendered. Admiral Badhwar has not only extensively interacted with all the Indian defence shipyards and many of the European shipyards, but has also represented Indian Navy/Government of India on numerous important national, bilateral and international committees associated with warship design and construction. Post retirement he has been the

Chief Operating Officer, Strategic Business with Pipavav Defence & Offshore Engineering Co, India's largest and the most modern Shipyard. For his distinguished services of very high order and outstanding contribution to the warship design and construction in the country, the flag officer was awarded 'VishishtSeva Medal' in 2002 and 'AtiVishishtSeva Medal' in 2009.

R.C. Bhargava joined the Indian Administrative Service in 1956, standing first in his batch. He worked in the States of Uttar Pradesh, J&K and then in the central government in Delhi. He was Agricultural Production Commissioner in J&K and joint secretary to government in the Power Ministry in Delhi. In 1979, Mr Bhargava was deputed to BHEL as Director Commercial. In 1981, he joined the newly established Maruti Udyog Limited. He was appointed as Managing Director in 1985 and became the CMD in 1990. He retired in 1997, but was re-inducted to the Board in 2003, when the Company was listed and Suzuki became a 54 per cent shareholder, and appointed Chairman in 2007. Mr Bhargava is widely credited for the success of Maruti, including introducing Japanese management practices, and building a vibrant component supply industry. He has received various awards, including the Order of the Rising Sun, Gold and Silver Star from the Emperor of Japan. He sits on the Boards of various well-known companies, as well as on the Advisory Boards of some foreign companies and has authored *The Maruti Story*.

R.S. Bhatia, a retired Colonel from the Indian Army, was formerly with L&T where he was instrumental in setting up and operating a greenfield plant for manufacture of defence equipment in a record time equipped with world-class facilities. He is a member of CII, National Defence Council and FICCI National Committee on Defence. He is currently Executive VP and CEO of Bharat Forge Ltd.

Vijaylakshmy K. Gupta is an officer of Indian Defence Accounts Service 1974 batch and is presently Member of Telecom Regulatory

Authority of India. She has had varied experience in core areas of administration and finance in Defence, Telecom and Social Sector in a career spanning over 37 years. Recipient of Colonel PyareLal Gold medal for the Best Thesis on Defence and Industry Interface in R&D and Production for Achieving Self-Reliance in Military Hardware/Software in the 39th National Defence College Course, she retired as Secretary Defence Finance in Government of India. In the Ministry of Defence with an annual budget of over ₹200,000 crore, she oversaw the complete financial management. As Member (Finance), Telecom Commission, she chaired 3G spectrum auction considered as benchmark for transparent price discovery. As Additional Secretary in the Ministry of Women and Child Development, she formulated policies of high social and economic impact for distributing the development dividend for women and children. HerPhD thesis was on Human Right of Women—Legal & Constitutional Guarantees vis-à-vis Implementing Mechanisms. Dr Gupta says, 'The guiding principles that I have followed throughout my career in the Civil Service are integrity and unity of thought, word and deed'.

Nalin Jain heads GE's aviation business in the Indian subcontinent and Thailand. He has the strategic and P&L responsibility of the region, for the civil and military aviation business of GE. He joined GE Aviation in 2005 and has played a key role in various initiatives helping establish GE Aviation's footprint in India. He has had the opportunity to work on various aviation-related projects ranging from setting up a maintenance, repair and overhaul, fractional ownership, training, low-cost airlines, leasing business, etc. Prior to his current position, he was Director Global Partnerships with Bombardier Inc. where he focused on business development and strategy in India and SE Asia. He is a mechanical engineer with an MBA from Indian School of Business. He started his career with French engineered materials conglomerate Saint-Gobain occupying different positions of increasing responsibility in sales and marketing, business development and general management.

Thelakat Jayadevan is with Thales International as the Director-Air Systems and the Key Account Manager for the Indian Air Force and the DRDO. He is a graduate of the Indian Institute of Technology, Chennai (BTech) and the Indian Institute of Science, Bangalore (ME). Mr Jayadevan has been with the Thales Group (known earlier as Thomson-CSF), initially in India as a Project Engineer and then as Deputy to the Country Head. He moved to France in 1990 and has been in several key positions in Export Sales and Marketing. He was assigned to India in 2007 as the Country Division Director for the Air Systems Division. He has since been closely involved with several major contracts with the Indian Air Force and the DRDO, including transfer of technology and offsets.

Shobhana Joshi joined Indian Defence Accounts Service in 1979. She is a Postgraduate in English from University of Delhi. She is an alumna of the prestigious National Defence College, Delhi, and was awarded MPhil in Strategic and Defence Studies. She has had a varied experience while serving in different capacities in the Government of India, both in defence and civilian organizations/ Ministries. Some of the assignments include Deputy Secretary (Ministry of Human Resource Development), Director (Finance), Ministry of Defence, Integrated Financial Adviser (Western Command), Army, Finance Manager and Joint Secretary (Land Systems) (Acquisition Wing), Ministry of Defence, Joint Secretary and Additional Financial Advisor (Ministry of Defence/Finance). Presently she is Additional Secretary and Financial Adviser (Acquisition) and Member Defence Procurement Board dealing with public financial management including budgeting, resource planning, expenditure monitoring and financial analysis of the capital acquisition budget of Ministry of Defence and also provides financial advice on acquisition proposals of the armed forces and related policy formulation and aspects of public procurement. Ms Joshi is on the visiting faculty of some of the premier training institutions in the government and has been giving lectures on 'Procurement Procedures, Contract Management and Budget & Financial Management' in the Defence Services.

K.V. Kuber, an alumnus of the National Defence Academy, and the Technical Staff College, specialized in Electronic Warfare. He commanded an Electronic Warfare Regiment in operations and has conducted EW operations in Western and Eastern theatres, both in conventional and insurgency environments. He has been an Instructor at the National Defence Academy in training potential officers. Col. Kuber founded and established the DOFA and was the chief architect of the offset policy since its inception in 2005. He has been closely associated with the formulation of the offset policy in 2006 and modifications to the offfset policy in 2008 of the Ministry of Defence. In his corporate life, Col. Kuber founded the Religare Strategic Advisory. He is presently an External Advisor with Ernst and Young (India), being a subject matter expert in DPP and Offsets, besides advising Indian companies on Defence and Homeland Security space. Col. Kuber is also an Advisor with the National Small Industries Corporation, a Mini-Ratna PSU of the Ministry of MSME, Government of India.

Arvind Lakshmikumar is the CEO of Serial Innovations. He is an established domain expert in applied imaging technologies, computer vision, robotics and intelligent systems. He has been a principal investigator and programme director for a number of international military and scientific programmes (DARPA, NASA, US Army, AFRL, NSF and DRDO) in the areas of vision and imaging and also serves on programme committees of major computer vision and robotics conferences. Dr Lakshmikumar has been a principal investigator and lead sub-contractor on multiple next-generation US military programmes, including Boeing-led Future Combat Systems and Combat Zones That See. He did his doctoral work in Robotics from Carnegie Mellon University and has Master's Degrees in Computer Science, Electronics and Chemistry from BITS Pilani. Prior to founding Serial Innovations, Arvind was the head of technology and operations for Sarnoff Corporation.

Vivek Lall, a renowned aerospace and defence expert, has held various key positions with his last stint being as CEO and President,

Reliance New Ventures. Dr Lall was heading the Boeing (Defence and Space) operations in India from 2007 till April 2011. Dr Lall has now joined the US-based global defense and nuclear giant General Atomics as the global Chief Executive for International Commercial Strategic Development for General Atomics Electromagnetic Systems. He is also affiliated with the United Nations in New York to advise on Broadband and Cyber Security issues for unique challenges within the global community and provide services that will help address them.

S.S. Mehta (Lt. Gen.), PVSM, AVSM, VSM, in a distinguished career spanning 41 years, has had the distinction of leading a Tank Squadron to Dhaka in 1971, earning him a Mention-in-Dispatches. He has held a number of command and staff appointments including ADG Military Operations, DCOAS (Planning and Systems) as well as General Officer Commanding in Chief of ARTRAC and Western Command. Postretirement, he has been Member of National Security Advisory Board and Director General, Confederation Indian Industry. He is a trustee of the Tribune Trust and a member of the Indian Council of World Affairs.

S.N. Misra served with distinction as Director, DRDO and Joint Secretary, Aerospace in the Defense Ministry before joining KiiT University as a Professor in Economics and Law. His book *Impact of Defence Offsets on Military Industry Capability and Self Reliance* is a pioneering work on the subject.

Besides Defence Economics, he is presently seriously engaged in issues such as Development Economics and Policy Issues in Higher Education. He is an avid contributor to prestigious journals such as *British Journal of Education, World Journal of Education* and *Indian Defence Review.*

E.S.L. Narasimhan is the Governor of Andhra Pradesh and Telangana. He was born in Tamil Nadu in 1945 and is a Gold Medallist from the Madras Presidency College. He is also a graduate in Law from the Madras Law College. Shri Narasimhan joined the Indian Police Service (IPS) in 1968, and was allotted Andhra Pradesh cadre. He moved to the Intelligence Bureau in 1972. In the Intelligence Bureau, Shri Narasimhan held various assignments relating to national/internal security and was the Director, Intelligence Bureau from January 2005 to December 2006. Shri Narasimhan has a keen interest in application of technology for national security purposes. He has done two tenures in the Ministry of External Affairs. In the first instance, he was posted as First Secretary in the Embassy of India, Moscow from 1981 to 1984. He has also served in the Ministry of External Affairs from 1996 to 1999 when he held the responsibility of overseeing the security aspects both human and physical of all the Indian Missions overseas. On demitting office as Director, Intelligence Bureau on 31 December 2006, Shri Narasimhan took over as Governor of Chhattisgarh on 25 January 2007 and moved over as Governor of Andhra Pradesh in January 2010. He also assumed office as the Governor of Telangana on 2 June 2014. He is an alumnus of the National Defence College.

Smita Purushottam is India's Ambassador to Venezuela. Working earlier at the Institute of Defence Studies & Analyses, she focused on how to construct a high-tech innovation ecosystem in India, the 'Reset' and Eurasia. Ms Purushottam has been published widely in reputed journals, books and newspapers. She headed the IDSA Eurasia cluster and the High-Tech Defence Innovation Forum which provides a platform for discussion on developing a high-tech manufacturing sector in India. She served in the Ministry of External Affairs, the Integrated Defence Staff Headquarters in the Ministry of Defence, India's Missions in Berlin, London, Beijing and Brussels, and Moscow. She prepared a paper 'Can India Overtake China?' at Harvard (2001), in which she recommended building a strong manufacturing sector. Ms Purushottam won a Fellowship for a PhD at Cornell University in 1980, but joined the Indian Foreign Service instead. Ms Purushottam earned a Master's degree (1st class) in History from Delhi University in 1979.

She taught History briefly at Jesus & Mary College, Delhi University. She speaks Russian and limited French.

Mrinal Suman (Retd. Major General) heads the Defence Technical Assessment and Advisory Service Group (DTAAS) of the Confederation of Indian Industry. He has also been conducting and directing highly acclaimed Defence Acquisition Management Courses for Indian and foreign industry, both in India and abroad. He superannuated in 2003 and since then has been actively involved in the promotion of defence–industry partnership. He is regularly invited to address, both in India and abroad, various chambers, associations and industrial delegations on various facets of defence procurement policies, offsets and business opportunities existing in the defence sector. Today, he is considered to be the foremost expert on myriad aspects of India's defence procurement regime and is regularly consulted with regard to proposed reforms. Major General Mrinal Suman is a prolific writer. His articles are regularly published in a large number of journals and have been translated in many languages.

V. Sumantran is Chairman of Celeris Technologies, a strategic advisory engaged in the domains of mobility, aerospace, defence and technologies. He is also a Distinguished Visiting Professor of the Indian Institute of Technology, Madras. He is an advisor to several leading Fortune 100 organizations in autos, industrial equipment and aerospace. Until recently, Dr Sumantran served as Executive Vice-Chairman of Hinduja Automotive, UK, the auto and manufacturing sector holding company of the Hinduja Group as well as Vice-Chairman of Ashok Leyland, India's 2nd largest Commercial Vehicle Manufacturer. For their Joint Ventures, he was the Chairman of Ashok Leyland Nissan Ltd, Chairman of Ashok Leyland John Deere Ltd and Chairman of Ashok Leyland Defence Systems Ltd. Dr Sumantran is closely involved in national bodies in India, engaged in Policy and Technology, and serves on the Scientific Advisory Committee to the Cabinet of the Indian Government and also served on the Science Advisory Council of the Prime Minister of India. He is a member of the National Manufacturing Competitiveness Council and served as Chairman, National Defence Council of the Confederation of Indian Industry until 2013.

George B. White is Director, South Asia International Strategic Partnerships, Boeing Defence, Space & Security (BDS) and serves as Director of Enterprise Industrial Participation for Boeing India. White is responsible for coordination across BDS and Boeing Commercial Airplanes (BCA) on Indian industrial participation programmes. In this role, he works to identify and leverage enterprise synergies; maintains relationships with the Ministry of Defence and Indian armed services customers to ensure successful execution of BDS industrial participation commitments; and identifies and develops Indian industry alliances to support industrial participation and growth for BDS.

Index